WHALE HUNT

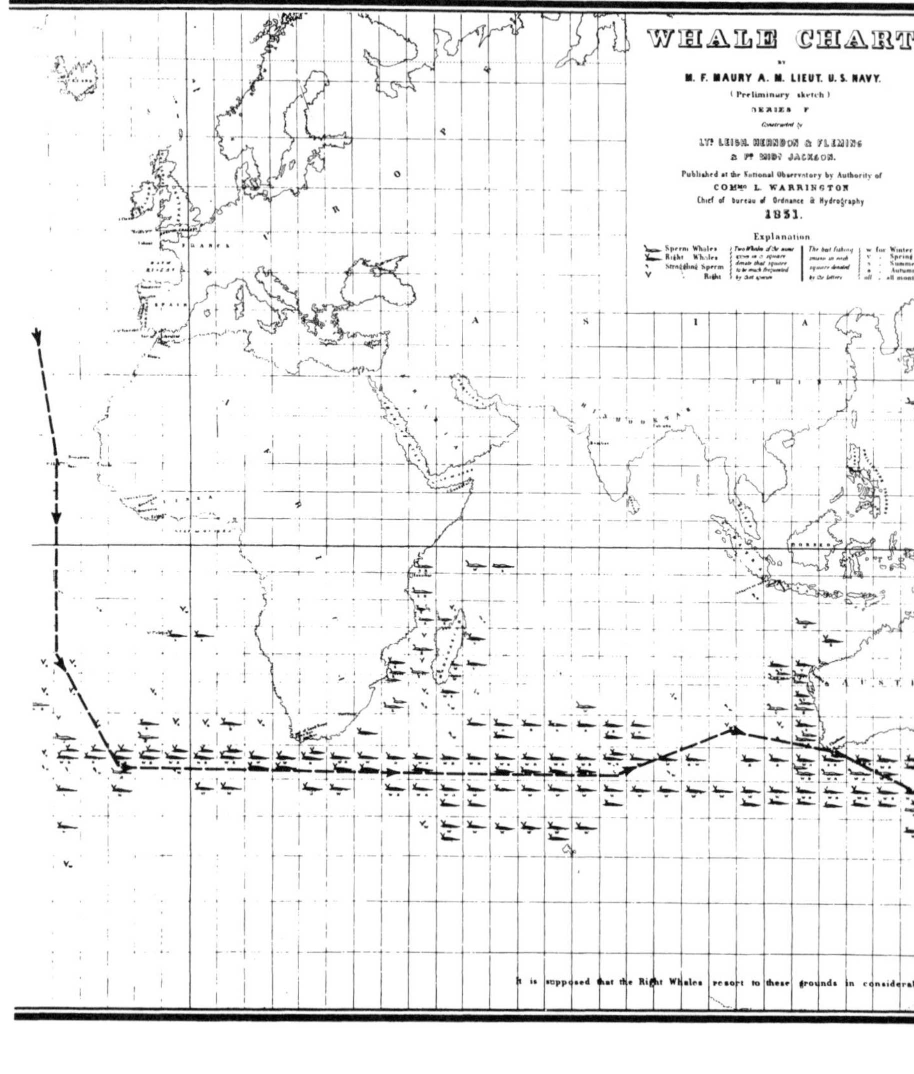
WHALE CHART
BY
M. F. MAURY A. M. LIEUT. U. S. NAVY.
(Preliminary sketch)
SERIES F
Constructed by
LT. LEIGH. HERNDON & FLEMING
& PT. MIDT. JACKSON.
Published at the National Observatory by Authority of
COMMO. L. WARRINGTON
Chief of bureau of Ordnance & Hydrography
1851.
Explanation
Sperm Whales
Right Whales
Straggling Sperm
Right
Two Whales of the same species in a square denote that square to be much frequented by that species
The best fishing season in each square denoted by the letters
w for Winter
v . Spring
s . Summer
a . Autumn
all . all mont
A S I A
CHINA
It is supposed that the Right Whales resort to these grounds in considera

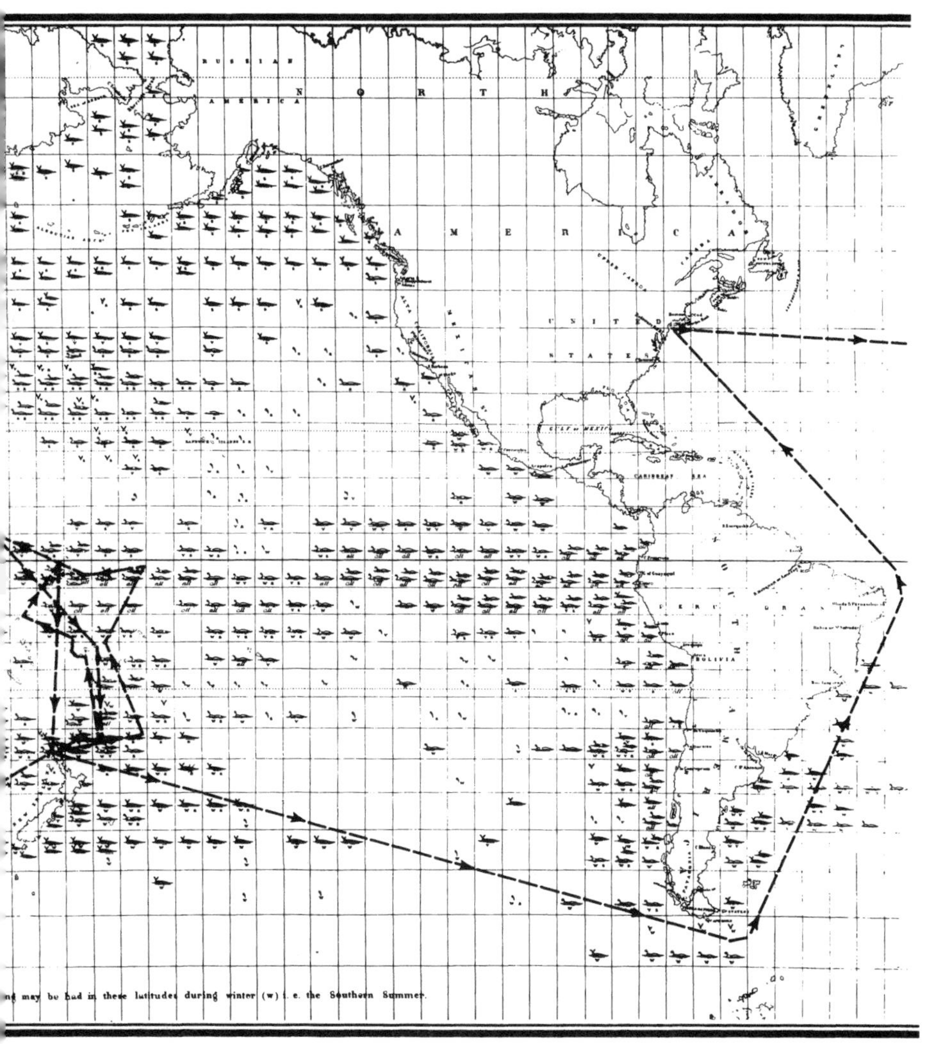

The approximate track of the 1849–53 voyage of the *Charles W. Morgan* is here superimposed on Lieutenant Matthew Fontaine Maury's whale chart, published by the U.S. Naval Observatory and Hydrographical Office in 1851.

Nelson Cole Haley (1832–1900) and his wife, Charlotte Brown Haley, ca. 1864. (M.S.M. 48.52)

WHALE HUNT

the narrative of a voyage by

Nelson Cole Haley

Harpooner in the Ship *Charles W. Morgan* 1849-1853

MYSTIC SEAPORT MUSEUM, INC.
Mystic, Connecticut
2013

Mystic Seaport — The Museum of America and the Sea — is the nation's leading museum presenting the American experience from a maritime perspective. Located along the banks of the historic Mystic River in Mystic, Connecticut, the Museum houses extensive collections representing the material culture of maritime America, and offers educational programs from pre-school to post-graduate.

Third Edition
Second Printing 2002
Third Printing 2013
(Originally published by Ives Washburn, Inc., 1948)

Cataloging in Publication Data

Haley, Nelson Cole, 1832-1900
Whale Hunt: the narrative of a voyage by
Nelson Cole Haley, harpooner in the ship
Charles W. Morgan, 1849-1853
Mystic, Conn., Mystic Seaport Museum, Inc., 1990
304 p. illus. 22 cm.
1. Whaling. 2. Charles W. Morgan (whaleship).
I. Title.
G545.H2.7 1990
ISBN 0-913372-52-8

Printed by Thomson-Shore, Inc., Dexter, Michigan

Contents

Foreword

UPON THE 150th anniversary of the launching of the whaleship *Charles W. Morgan* and the 50th anniversary of her arrival at Mystic Seaport, we are pleased to publish this commemorative edition of Nelson Cole Haley's exciting memoir of his voyage aboard the *Morgan*. It is fitting indeed that Mystic Seaport Museum publishes this volume for, like the vessel herself, the original manuscript of Haley's narrative is preserved at Mystic.

The *Charles W. Morgan,* America's last surviving wooden, sailing whaleship, was not simply the vehicle for Haley's account; she was also the spur that brought it first to publication in 1948. Haley wrote this draft of his narrative in the 1890s. It is presumed, but not established as fact, that he based his account on a diary or journal kept during the voyage. Until 1944 the manuscript never left the possession of Captain Haley's family, except for a brief period when a copy was made by the History Department of the University of Hawaii.

In 1941, when the *Charles W. Morgan* was brought to Mystic for preservation, newspaper accounts of this move aroused the interest of Captain Haley's heirs and led to correspondence between Captain Haley's daughter, Mrs. May

Rothwell of Honolulu, and Carl C. Cutler, Managing Director of the Marine Historical Association, now Mystic Seaport Museum. The enthusiasm with which a typewritten copy of the narrative was received by the Association prompted Mrs. Rothwell to present the original to the Museum in 1944.

Ives Washburn first published the work in 1948, ninety-nine years after Haley set forth aboard the *Morgan*. A second edition came out in 1967. Due to continuing strong interest in the work, Mystic Seaport Museum now issues this third edition, bringing to a new generation this firsthand glimpse of life aboard a classic American whaleship.

Even as public opinion has become firm in its opposition to the continued killing of whales, the fascination with the history of whaling remains strong. In this Haley does not disappoint. There is plenty of the primal struggle between man and the great leviathan here.

But there is more, too. The clash of cultures as western whalemen invaded the islands of the South Pacific is evident in Haley's descriptions. And, like a good sailor, Haley yarns for the reader, telling stories he has heard, such as the one about the young woman who disguised herself as a sailor aboard the Nantucket whaler *Christopher Mitchell* for eight months.

As a narrator, Haley gives us much of himself. He is not a moralizing reformer with a grudge against the industry in the style of Francis Allyn Olmsted (*Incidents of a Whaling Voyage*, 1841) or J. Ross Browne (*Etchings of a Whaling Cruise*, 1846). Neither is he a one-voyage whaleman who rushed to publish a travelogue based on his whaling cruise, for he remained a whaleman, off and on, into the 1860s. Nor does he resort to the formal literary conventions of his time, which would have made this a much less spontaneous and less complete view of the life of the whaleman.

At Mystic you can walk the deck of the *Morgan*, see the dark forecastle, even feel the worn spokes of the wheel. But it takes Haley to breathe life into the men who once inhabited this ship. Go aboard after reading his narrative and you'll feel them all around you.

— J. REVELL CARR
President & Director
Mystic Seaport Museum

Note of Introduction

NELSON Cole Haley was born on 7 March 1832 at New Bedford, Massachusetts. His father, John Haley, died soon after, leaving his widow, Meribah, daughter of Otis and Mercy Russell, to bring up the boy and his two sisters, the eldest child being then five years of age. As Nelson's boyhood fell within the years when New Bedford was completely engrossed in whaling, it was natural that early in life he should form the resolution to become a whaleman. Unable, however, to obtain his mother's permission, he ran away, as was also natural, at the age of twelve, and shipped in the whaleship *John*.

His first cruise ended shortly after his sixteenth birthday. The winter of 1848–49 he spent in school, at his mother's request; she having meanwhile remarried. But within a few months he was off again, this time in the ship *Charles W. Morgan*, Captain John D. Samson, on the four-year voyage that is the subject of this narrative.

After thirty years as a whaleman, including at least ten as master, Captain Samson was a good judge of young whalemen. Young though Haley was, Captain Samson evidently had good reason to feel confidence in the lad, for he shipped him on in the responsible capacity of boatsteerer, or forwardmost oarsman of a whaleboat, whose duty it was to dart the harpoons into the whale, then steer the boat while the mate prepared to kill the whale with a lance. As Haley tells us, he already had been practicing on sharks and dolphins; and after safe passage around the Cape of Good Hope and through the Indian Ocean, he struck his first whale in the South Pacific when he was seventeen.

Returning to New Bedford in the summer of 1853, he heard news, from the Sound pilot off Block Island, of schoolmates who had come back with pockets jingling from the California gold rush. His long cruise round the world, so he presently found at the owner's accounting, had paid him an even $200.00, after expenses. In a family reunion at his stepfather's home in Portland, Maine, to use his own words, "A capstan-head consultation being held by these different boat-headers, including the Colonel and Mother, it was determined that yours truly should abandon the sea and go West to grow up with the country."

He set out for the little frontier town of St. Paul, going by way of Lake Erie ("How much alike the water seemed! It was hard for me to believe myself on a vast body of fresh water miles inland from the mighty salt oceans that had been my home"), and so from Galena up the Mississippi. During the time he remained in Minnesota, to the satisfaction of his employers he managed first a general store, and then a large lumbering camp and sawmill.

But, "during the time I had been in St. Paul, numbers of letters reached me from New Bedford, wanting me to go a whaling again. The offers were quite flattering. This no doubt had some effect on my waywardness." When he announced to his employers that he still felt the call of the sea, they offered him inducements to stay on. "Among the offers," he says, "were two blocks of land near the Falls of St. Anthony. I think this was in the woods about eight or ten miles from St. Paul. It was of no use. The only answer I could give them was 'Kismet,' as the reason of my going."

Haley left Minnesota in early winter, just before the Mississippi froze, to return to New Bedford. Having decided to try right whaling in the North Pacific, he reportedly signed aboard one of the last two vessels set to depart that season. However, surviving records do not offer any evidence of his presence in the whaling fleet until July 1857, when he signed aboard the ship *Metacom* as first mate for a voyage to the North Pacific. The *Metacom* was lost in December 1860, after three season in northern waters.

Haley must have left the *Metacom* after a season and returned home in another vessel, for in August 1859 he departed New Bedford as first mate of the ship *Navy*, again for a North Pacific voyage. She made a very productive four-and-a-half-year voyage under Captain Andrew S. Sarvent, but, again, Haley did not complete the voyage. He was discharged at Hilo, Hawaii, on 11 November 1861. According to family tradition, he then served as an officer in the Hawaiian whaling fleet.

On 17 March 1864, he married Charlotte Brown, daughter of Robert Brown, a retired whaling captain from Groton, Connecticut, who had settled in Honolulu during the early 1850s. Six children, Ellen, May, Frederick, Charlotte, Ann, and Edith were born to them.

One of his daughters, Mrs. May Rothwell, gave this description of her father:

He was not a large man, being only five feet, six and one-half inches in height; yet he did not appear small. His shoulders were broad and he had the strong, well-muscled frame of the active outdoor man. His deep-blue eyes could be merry or serious, or, occasionally, severe. His golden-brown hair, in middle age, became snow white, but his full beard remained dark red until his death. Although not a well-educated man, he was highly intelligent and very well informed. He had a keen sense of humor, but it was always kindly, though sometimes of a quizzical quality our friends could not always appreciate. His fund of anecdotes, especially of whaling experiences, seemed to be inexhaustible. He was a kind but strict disciplinarian. His servants and plantation workers were devoted to him. He loved flowers and animals, and, like all sailors, was exceedingly neat. Honest above the average, he trusted all men. When misfortunes came he bore them uncomplainingly and quickly rose to become his own merry self. He loved his children and adored his wife. Quick to anger, he never held resentment.

For several years following his marriage, Captain Haley had engaged in the export of sandalwood to China, and of pulu, a red cotton from the native Hawaiian fern tree, to San Francisco. When these products were exhausted, he turned his enterprise to sugar planting. Because this culture was then imperfectly understood, and seed of the varieties suitable to local condition was not available, the attempt was unsuccessful. Following this venture, Captain Haley removed to Seattle, where he engaged in business and was at one time assistant postmaster. Always active and of a venturesome disposition, he finally embarked upon the supplying of food and equipment to the Alaskan miners in the gold rush of 1897. He was stricken with pneumonia and died in the late winter of 1900, at Sheep Camp Hospital, in Alaska.

OF HIS wealth of whaling anecdotes, as mentioned by his daughter, Nelson Cole Haley has left in written form the continuous narrative that follows. Set down in the leisure of some later time, it is in no sense a mere log of the whaleship's voyage. It is the full story of how the *Charles W. Morgan* was handled during her cruise, 1849–53, and of Haley's own adventures aboard ship and among the islanders of Oceania. On the lighter side, it accounts for the moments of respite, in the long run scarcely less hazardous than those of the whale hunt, that were enjoyed by crewmen on liberty when the ship put in for fresh water and to recruit supplies from the natives of far lagoons. At such times an ill-advised choice among neighboring coves or atolls could mean the difference between relaxing at a hospitable roast of fish and taro or furnishing the native diet with a welcome windfall of long pig, for sudden massacre on the coral beaches was then a common occurrence. Commoner still, after dalliance beneath the mangrove shade of the friendlier isles, was lingering disability.

Perhaps because of his youth, although more likely because Haley had an eye on the prospect of one day qualifying for the quarterdeck, he was disposed to take a clinical view of the delights that awaited Jack ashore with a pocketful of trinkets (fishhooks, mainly) to trade with Kanaka maidens. Remarks the young boatsteerer about the kava-chewing *coryphées* of Strong Island (now Kusaie in the Carolines) "that hell's kitchen," when they stood up to dance, "As most of the time during this exhibition the women faced us backwards, a good chance was afforded for us to see the full development of their muscles." But more especially, of course, does he have to tell of the "close chances" he experienced while at sea: the opportunities to strike sperm whales, only a wave or two away from the bow of the boat headed by the *Morgan*'s daredevil second mate, Mr. Griffin.

The New England whaling industry was then at its peak. During the very time these adventures occurred, Herman Melville, home from like experiences in South Pacific waters, was engaged in the composition of *Moby-Dick*. As will be seen, Haley's Captain Samson was not another Captain Ahab. Although he was, in Haley's opinion, as strict a man as sailed out of New Bedford, ships under his command were asked to engage in no metaphysical quest. Oil was the object. When a day-old calf whale stove one of the ship's boats and nudged up to a second, mistaking them for its mother, the penalty of inexperience was an iron in its hide. "The poor little thing," Haley calls it; but when it had been hoisted aboard by the flukes, whole and entire, "He made," he adds, "about two barrels of oil. . . . The little whale gave us boatsteerers a fine lesson in regard to the anatomy of its species, and was to me a great help in reaching the life of many a sperm whale later on."

How well Haley succeeded in mastering the details of his laborious and dangersome trade may be seen at once from the passages here following. These have been lifted from the body of the text in order to furnish an introduction, in the author's words, to the nature of the whaleman's monstrous prey as he saw it, and to the hard-learned, lethal economy of that lapstraked cockleshell, the cedar whaleboat, which was lowered away upon the mile-deep waters with a crew of six men for the pursuit and kill.

The eye of a 100-barrel whale is about the size of that of an ox, but it must be very powerful, for at times, when alarmed, he has the power to detect danger from long distances. The ear perhaps is the more wonderful organ of the two, as many instances have occurred in my experience when whales have become alarmed from sounds two or three miles away.

A person who does not know where to look for it might hunt a long time before he finds the ear from the outside. It is a little behind the eye, and the opening is so small that one can hardly insert the end of the little finger in the external opening; but on cutting through the blubber and following its passage to the brain, there can be found an increase as it extends, like the tube of a trumpet on a small scale.

The mouth of the sperm whale is a wonderful-looking affair, being covered, from the end of his long jaw to as far as one can see down his throat (tongue, which is quite small, and all), with a shining white membrane, like satin. The teeth, however, do not look so pretty.

The eye and ear of the Balena whale do not differ much in size or formation from the sperm whale's, but there all similarity ends. The head of a right whale or a polar whale is a horrid thing to look at; and more so when he is coming head-on towards you, scooping (feeding).

Remove from his head the two immense lips that in a 250-barrel whale will measure some twenty-five feet in length and average eight or nine feet in width, with a mean thickness of, say, fourteen inches, weighing two tons each. (The lips are fat but very firm, and seem composed of entirely different matter to any other part of the body. This can be cut with sharp spades and knives without much trouble, and the oil comes out of the scraps freely). You will then have a crooked upper jaw some twenty-five feet long to which are attached the ribbed vertical scimitar-shaped slabs of whalebone, a hundred or more on each side.

The top of the head is a bone covered with a light membrane, and is of no use whatever, but it has to be hoisted on board, as there is no other way to save the whalebone that is fast to it. The slabs of whalebone are embedded by the butts, ten or twelve inches deep, in a grayish-colored substance called by some whalemen the gum. This substance when fresh will cut quite freely, but when dry it has somewhat the nature of the bone it holds in place.

After the head bone is on deck, one end of it is hoisted clear, and with a spade cuts are made in the gum at the roots of the whalebone and on the sides. After starting, the bone from its own weight will soon tear the gum clear from the scalp bone and come tumbling down on deck as the head bone is hoisted aloft. The whalebone is separated into junks of five or six slabs by cutting the gum between them; and this makes it convenient for stowing below, out of the way among the casks, until time can be found to cut each slab separate and scrape it clean of the gum. Such work is done when no whales are in sight.

The inside edge of each slab is fringed with hairy fibres, through which he strains the water and those intricacies that are his feed, which in right whales is called the brit, a minute substance almost colorless when alive; but when dead and floating on the surface of the ocean it has a reddish appearance, looking like dust scattered over the water. The feed of the polar whale has more substance, as it is composed of a water insect that looks, more than anything else, like a small spider with short legs. These have a maroon color.

The tongue of the Balena whale is very large. I have seen over twenty barrels of pure oil tried out from one tongue; they are one mass of fat. When the whale has scooped (as it is called by whalemen) enough feed into its mouth, the lips will be brought together with a snap. Closing at the same time his lower jaw, and with his tongue pressing hard against the slabs of bone held firmly in place by the lips, a stream of water will rush out at the corners next the body, through openings nature has formed for that purpose. These at times have the appearance of the jets seen coming from the sides of steamers.

The lower jaw is composed of two bones that extend from his body, one on each side, some thirty feet forward, where the ends are joined together by a tough cartilage. These have a shape like a round-pointed shovel. Where the bones, which are oval in shape, join the body, they are some three feet in circumference, tapering to about one foot where they join each other.

The lips are fast to this lower jaw, coming to a round point at the forward end. The after end of each lip is joined to the body by one corner; otherwise, they are loose, and when feeding these are dropped away from the part of the head that holds the bone, that part being raised with the ends of the slabs of bone sticking out. The lower jaw, when dropped enough to show an opening big enough to drive an ox team in, and he coming quickly towards a person for the first time, a vision of such horror is unfolded to the view that one might wish he had not come whaling. . . .

That one who is not familiar with the manner in which whale ships carry their boats may understand, I will try to describe it:

Each ship has from three to four boats that hang from wooden or iron davits, two boats (or three, as the case might be) on the larboard side, and one on the starboard quarter. These are sharp at both ends; and for speed, stability and buoyancy in riding seas, no boats in the world are their equal. They are twenty-four feet in length over all, and have seats for five oarsmen, a hole in the second thwart forward for a mast, and a platform at each end for steersman and harpooner.

Between the two after thwarts is set a tub in which is spirally coiled in concentric layers the softest of manila ropes, two-thirds of an inch in thickness, 220 fathoms long. One end of this is to make fast to an eye spliced in the strap that holds the iron to the pole. The other end is left hanging loose over the edge of the tub as a matter of safety, and to fasten a second line on, if needed, from another boat, should the whale sound deep enough to take out the entire length. (I have seen a large sperm whale take out four of these lines, one bent to the other, a mile in length, and this on an up-and-down sound; and get away.)

The end of the line that is to be made fast to the iron is first taken off from the tub under the loom of the after oar and passed around the loggerhead that each boat has in the stern. It is then carried forward over each oar, between the men as they alternately sit at the opposite gunwales. About ten or fifteen fathoms of it is coiled snugly into the head of the boat fitted for that purpose; and this is called the stray line, which is quickly thrown overboard after the iron has been darted into the whale, so as to allow the boat a chance to stern off him before the line might become taut by his rolling, as he sometimes does.

The line passes between the extreme ends of gunwales that are far enough apart to allow play to a brass roller with a groove in it, over which the line runs freely. It is held in place there by a pin of tough wood pushed through holes in the ends of the gunwales above. This end of the line being made fast to an iron, all is ready for the boat-steerer to dart.

First, though, you have to sight your whale; if not the hulking hump of a solitary bull sperm, then at any rate what

Haley is fond of calling "the low bushy spouts" that betray where the herded cetaceans are quietly basking, unaware of eager lookouts with eyes that, although scarcely more powerful than their own, have been sharpened by thought of profits to share and spend on Nantucket or in New Bedford or in New London, half a world and perhaps two years or more away. "'T-h-e-r-e s-h-e B-l-o-w-s!' was the pleasing sound from the masthead": and it is the keynote of Nelt Haley's narrative.

A word should be added about the editing of this text. It has been broken into chapters for the reader's convenience; and some passages have been cut, to bring the book down to manageable proportions. The text as printed, then, is not a rewrite, and nowhere has so much as an entire sentence been written in. It is all Haley himself. The script has been copy read for punctuation and routine spelling; cases and tenses (for the most part) have been straightened out; and hook-and-eye words have been supplied, where needed, to ease over the occasional bumpiness of a style that is the unaffected one of a congenial writer of letters home. As the reader will perceive, the author is no mean storyteller; and his material was held to the natural order of its interest by the shape of the voyage that gave it rise.

To illustrate the text we have selected lively pen-and-ink sketches by Robert Weir (1836–1905), which, like Haley's journal, are in the Museum's collections. The talented son of artist Robert W. Weir (drawing professor at the U.S. Military Academy at West Point), and older brother of artists John F. and Julian A. Weir, Robert Weir ran away to sea for some unexplained reason in 1855. Although Weir penned them to illustrate the journal of his voyage aboard the whaleship *Clara Bell* (1855–58), these sketches are so evocative of American whaling that they reflect Haley's experience aboard the *Morgan* equally well.

WHALE HUNT

Crew of the *Charles W. Morgan*, 1849–53

NAME	BIRTHPLACE	AGE	COMMENTS
Captain John D. Samson	New Bedford, MA	46	
1st Mate Thacher Packard	Sandwich, MA	30	
2nd Mate William Griffin	Waterloo, NY	29	
3rd Mate Roland Briggs	Mattapoisett, MA	25	
Stephen Barnum	Alford, MA	24	
Thomas A. Tyler, Jr.	Norfolk, VA	20	
Nelson C. Haley	New Bedford, MA	18 *[sic]*	5'4½", light hair and complexion
Benjamin Butman	Marblehead, MA	26	
William L. Covell	New Bedford, MA	16	
Charles Coy	Lowell, MA	25	
Moses H. Gardner	New York	22	
Enos Eastwood	Cicero, NY	23	deserted at Rotumah
Daniel M. Sampson	Middleboro, MA	15	
Lewis Dorr	Foreign		deserted at Eoa (Tonga)
Amasa Colby	Eaton, NH	20	deserted at Horn Island
Smith Tucker	Worcester, MA	21	
Benjamin Olney	Newport, RI	23	
Francis Warner	Foreign		deserted at Rotumah
John Barker	South Pacific		discharged at Rotumah
Ben Rotch	Foreign		discharged at Fayal
William Auza	Foreign		
Joseph Whitman	Scituate, MA	18	
Matthew Gilgannon	Rochester, MA	21	
Thomas Phillips	Foreign		
James W. Parker	Foreign		deserted at Rotumah
Martin Whitehead	Pottsdam, NY	21	deserted at Fayal
George Duffy	New Bedford, MA		
Vincent D. Crowl	Smithfield, RI	21	deserted at Rotumah
David Van Riper	New York		discharged at Strong (Kusaie) Island
John Adams	Rotumah		discharged at Strong (Kusaie) Island
Joseph Maxwell	Rotumah		discharged at Strong (Kusaie) Island
Thomas Ryan	Lancashire, England		deserted at Bay of Islands

Compiled from crew lists on file at the New Bedford Free Public Library, with assistance from Paul Cyr, and with reference to John F. Leavitt, *The Charles W. Morgan* (Mystic: Mystic Seaport Museum, 1973).

To the Indian Ocean

AFTER stopping in New Bedford a short time, visiting our friends, when I arrived home in the ship *John* from my first voyage, we went to Portland, Maine, where my mother's husband lived. I stayed at home and went to school, part of the time. I did not have much wish to try the sea again until a Captain of our acquaintance came on a visit to our house and wished very much for me to go with him on his next voyage. Much against my mother's wish, I consented to do so.

A few months after, Captain Sampson sent me a letter stating that he would sail at a certain time, and for me to come on and join the ship, which was one of the best that sailed from the port. I left home; and on my arrival in New Bedford, I found it as he said about the ship. She was the *Charles W. Morgan,* one of the crack ships, and belonged to Edward M. Robinson.

The Captain went into Mr. Robinson's office with me, when I went to sign the shipping papers. I found him to be a tall man (six feet at least) with keen black eyes and a hawkbill nose, with a very dark complexion. I then saw why he was nicknamed "Black Hawk." He arose and shook hands with the Captain; and looking down on me with his eagle eyes, he said, "The ship you want to sail in can command the very smartest men for officers and boat-steerers that the city affords. You look young and small for the position you would have to fill on board that ship."

I felt a little embarrassed (I was but seventeen years old, and stood but little over five feet in height) and hardly knew what to say. Turning to the Captain, he said, "You pick your own men, and if he will suit you for a boat-steerer it is all right." The Captain told him it was. Then he turned to me again and said, "Do you think you could strike a whale?"

I told him I had struck sharks and dolphins; and that whales were so much larger, I thought I could if the boat I steered got near enough to one. He looked at me a minute with a twinkling eye and replied, "Well, often valuable articles are in small parcels."

I signed the ship's articles and went out.

On Sunday the 3rd of June, 1849, we went on board the ship. She was out in the stream at anchor. The fog was so thick that we did not get under way then. About midday it cleared off. About the same time, though, the Captain's mother died and the news came off to him, so he went on shore to see her buried. The sails which had been loosened were furled and the ropes coiled up, and orders were, "No one allowed on shore." The officers employed the chance to prepare the boats for whaling, and to pick out the boat-steerers and watches. I was chosen by the 2d Mate to steer his boat; and of course that made me in the Starboard watch.

The manner of choosing watches and boat-steerers on board whale ships is this: all hands are called aft on the quarter-deck, the men on one side, the boat-steerers abaft the capstan. The Mate chooses his boat-steerer first, 2d Mate next, 3d Mate next, and 4th Mate next. (If there is a 5th

Mate, he of course has to take the one that is left!) The Chief Mate then picks out one man from the crew for the Larboard watch, then the 2d Mate picks one for the Starboard watch, then the 3d Mate picks another, and so on until no more men are left. If there is an odd man, he goes into the Larboard watch.

During the two days we lay waiting, the boats were put in complete order for catching whales: harpoons, lances and lines were put in the boats, and everything was made shipshape. This always has to be done as soon as the ship is at sea; and in our case we had the best chance possible to do it, with no sails to trim or anything to call attention but that.

On the morning of the third day after making our start, we got under way with a pleasant breeze from the N.W. As our course out of the bay was about South, it brought the wind on the quarter, so every sail stood out full from yards and leeches. We went by the lighthouse on Clark's Point flying, the old ship carrying a white bone in her teeth.

I took a look at the lighthouse as we left it behind, and the thought came into my mind, How long will it be before I will see it again? Perhaps this is my last look at its white towering sides and network of iron surrounding the lanthern on the top. It went through my mind, Why should I cause myself such sad feelings by taking this voyage? Here I was, leaving home perhaps never to return, and for no satisfactory reason I could give; leaving behind a happy home and the friends who had done what they could to have me stay with them. I knew what I had to face, that at least was sure: storms, gales, hurricanes, lee shores, and whales' jaws and flukes. For what? Not for money! Because not much of that comes to the crew, and but little more to boatsteerers. Well, it might be for the wish to command a ship, in proper time. Still, was it worth the candle?

Soon, though, my mind was taken up with the duties I had to attend about the ship. The Pilot left and we filled away the mainyard on our course E. by S. with all square sails set alow and aloft, all hands at work, some storing the anchors, some tauting up riggin', others stowing away spare spars and lashing up loose matter for a long ocean voyage.

Favoring winds followed us across the Atlantic. No whales were sighted. We lowered the boats to practice the men in rowing and the use of oars at every opportunity; and one day a large school of black fish came in sight. The boats were lowered and went in chase of them. We captured three or four, which made, when the blubber from them was tried out, about five barrels of oil. These fish, so called, are a variety of the whale, but not like any other except the sperm whale, and only then as regards the oil, which is something like sperm but not quite as good. However, it is the only kind of oil that will mix with sperm in small quantities and not be detected.

After cruising about here and there, we shaped our course for the Azores, or, as called by whalemen, the Western Islands, which in a few days we sighted. We had seen no land since leaving home some six weeks ago. The green hands had recovered from their seasickness, and with eager eyes they watched the approach of the ship as it came nearer and nearer to what seemed to them a garden in the ocean, as different spots of cultivation showed all the colors it was possible for vegetation to do. The land seemed to be under high cultivation from summit to shore except where the small towns were situated, which showed in gleaming whiteness, being built of stone and whitewash. We stood in towards Fayal, which is the principal town of the group.

Amongst our green hands was a lank Downeaster, so green that even the others made sport of him. He had been the most seasick one of the lot and had hardly recovered when we made land. He was in the Mate's watch and the other officers had been told to let him do about as he was a mind until such time as he got well, so he went on deck with his watch or not as suited him, and but little attention was paid to him. About the time we approached the land near enough to haul aback the mainyard and lower a boat for the Captain to go on shore to take the letters and buy some recruits, as we did not intend to drop the anchor here, who should come poking aft on the quarter-deck but our sick man. The way he was rigged up was stupendous. Such a sight would be seldom seen on a ship's deck, as a sailor. The men forward were choking with glee.

None of the officers caught sight of him until he was past the main riggin'. The first to do so was the 3d Mate, who was leaning over the capstan. He took one look at him and yelled, "Oh Holy Ghost!" clapped his hands to his sides and burst into shrieks of laughter, saying when he could, "I shall die, I know I shall."

By this time the other officers had joined in the chorus of laughter. The Captain, hearing the uproar, came on deck and caught sight of what to him at first was a stranger standing on the quarter-deck, with his officers and crew yelling with laughter. He hardly knew what to make of it, and was on the point of asking what it all meant when, taking another look at the strange figure, he saw who it was, and he then laughed as loud as the others; and well he might.

The hat he had on, an old bell-crown beaver, looked as if it had been white one day. The wrinkles had been straightened out of it as much perhaps as possible, but one side of the rim could not be made to match the other, as it had been broken down. The coat was a blue claw-hammer with some brass buttons on it, his vest was bright scarlet, his shirt had been ironed but, owing to bad stowing or getting a little wet, one part of the standup collar hung limp on his coat over a red necktie. His pants, so tight that his legs looked like large-sized bean poles, came only halfway down from his knees; and with his heavy boots he looked the most startling comical transformation out of a sailor.

As soon as the Captain got charge of himself and could speak, he said to him, "What in the name of goodness do you mean by coming aft on the quarter-deck in such a rig as that you have on? Where did you get such an outfit?"

He replied, "These clothes and hat have been for a long time my Sunday ones, and the man who shipped me on board this ship told me I had best take them along, for when I went on shore on liberty I would need a different rig than that worn at other times, and to take good care of my hat. But when I was too sick to do anything, one of the boys got it out of my chest and sat on it, so it does not look as smooth as it used to do."

"Never mind telling anything more about the clothes, what next?" said the Captain.

"Well, you see I have about come to the conclusion, as I have not been much use to you, and do not think a sailor's life suits me one bit, I would just ask you to let me off here and I could walk back to the old state of Maine, not caring how far it was. That is how it stands, Captain."

The Mate told him to go forward and take off those things; that no doubt he might be able to wear them on shore if he wished to do so at some future time, as there would not be any chance to land him just then, and he could not be spared. He went slowly towards the forecastle looking quite downhearted, and shortly after came on deck dressed as the rest of the men were.

The Captain soon after went on shore, taking two other boats with him to load with fruit, onions, potatoes, and chickens. During the day the boats made numbers of trips off and on from shore to ship. At sundown the Captain came on board, we squared the yards and shaped our course for another island called Saint Nicholas. At daylight it was a short distance off. We spoke two or three ships that were lying off and on the town—one of which was commanded by our Captain's brother. They had not met each other for sixteen years, and were much pleased to do so. If that chance meeting had not taken place, they could never have met in this world again, as the brother returned home with his ship and died a few days after.

We did not require much more in the way of recruits; but now we went on shore and the Captain sent off a boat-load of oranges, and some more eggs and chickens, all of which were very cheap here: eggs 5 cts doz, chickens $1.00 doz, oranges 50 cts per 100. I went in charge of the Captain's boat, the other boat being in charge of the 3d Mate.

When I landed with the Captain in the morning, he told me that I could let the men leave the boat until noon but to have them on hand at that time. I told the boys they could ramble about the town until then, so they went off to some of the many places where wine is sold very cheap, about 5 cts a quart.

The city, being built mostly of stone and everything on the outside whitewashed, at a distance gave it a fine look. The white shone out finely in the sunlight, but on close inspection the effect was destroyed to a great extent, for dirt could be seen on most of the walls as though it had been put there to make them as dirty as the inhabitants. Still, I think the walls were the cleaner of the two.

As it was getting along towards noon and time to meet the Captain, I steered towards the place of meeting. The men were all there and shortly the Captain appeared. He told me to take the men some place and give them dinner, and handed me $1.50, saying that after dinner I was to find the 3d Mate and go on board ship when he did. As for himself, he would go off in one of the merchant's boats that he had business with.

We started off for dinner, came to a place that had the cleanest looking appearance, and went in. I made known our wants more by signs than otherwise. The Don who ran this place was a distinguished looking person, and with the grace of a Spanish grandee set out a bottle of wine with glasses for each of us. We all took a tot, as an appetizer; shortly after, he informed us dinner was ready. The table had quite a clean cloth on it, the tumblers and plates also looked pretty clean for a Porta'gee place, a bottle of wine was at each plate, and a brimming platter of chicken stew in the center. With fruit and flowers placed here and there, it made a display almost too grand for a common sailor to sit down to. We helped ourselves to chicken and commenced to eat.

First one and then another stopped at the first mouthful, laid down knives and forks, and like myself almost spit out what we had in our mouths. Some of the boys said, "What in hell is the matter with this stuff?" I knew at once what the trouble was. The one who had cooked it had stuffed it full of garlic, and to us all, if rotten eggs had been in it, the taste no doubt could not have been much worse. We all left the table and went in the front shop. I had hard work to make our friend Don Ceaser Diablo understand he must cook us something that had no garlic in it. He seemed astonished, but rushed off into the den in the rear, and soon we could hear a big powwow going on out back. He shortly returned, somewhat excited, but gave us to understand he would soon have something for us.

In the meantime, he came pretty near to getting the boys so full of wine that when we sat down to the table again it was a question whether some of them could have told whether what they were eating had garlic in it or not. I found the last lot not so bad but what I could worry it down. It would be hard to have anything cooked in utensils used by them that would not taste of garlic. The fruits were nice, more particularly the ripe figs, so I made out a hearty meal.

When I went to settle the bill, the Don pulled out a

strip of yellow paper about six inches wide and at least a yard long, with items on the whole length of it. I fairly grew pale and thought, How in the name of goodness am I going to pay all that bill with the money I have? When the Captain handed me the $1.50 he said, "That will be enough to get the men's dinner with and give them a drink besides, as things of that kind are very cheap here." But here was a bill, by the looks of it, for five or six dollars. I had a little money outside of what the Captain gave me, but I was sure it would not be enough.

Of course it was impossible for me to tell how much it amounted to when I was shown the figures, as the bill was in Portuguese money. I put on the counter about $3.50. The Don's eyes stuck out like a crab's, and his lower jaw wobbled as if it were unhung. He let off a shot at me in Portuguese, sorted over the money on the counter, picked out 75 cents and shoved the rest back towards me. To say I was dumbfounded would hardly express my astonishment at the small amount charged for the food and wine we had stowed under our hatches.

It was with feelings of relief that I put the balance of money back into my pocket and walked out into the street. As it was about time for me to hunt up the 3d Mate, we started on our way to the mole where our boat lay. The other boat was there, and the men who were in it said, "The 3d Mate told us to keep in it, all ready to shove off when he comes back."

I went up on the principal street and stood a few minutes looking up and down. Soon I caught sight of him and the 2d Mate of another ship that also was lying off and on the port. As they approached me, I could see they had been taking in more wine ballast than they needed. When they stopped in front of me, I told the 3d Mate what the Captain had said about my finding him and going aboard when he did. "All right," he replied, "let us take a bit of a spin around the town before we go off." I thought he had been spinning plenty enough then, but he was above me in rank and so we started on. Shortly after, we went into a shop that kept for sale dry goods, shoes, clothing and all sorts of knickknacks. A counter ran nearly the whole length of one side and across the end. What they wanted in there I failed to see.

When we went into the shop, three or four dapper young Portuguese clerks behind the counters bowed, smiled, and put on the air of being so pleased to see us that it seemed to rouse the anger of the officer who accompanied us. He yelled at them, in a voice like hailing from deck in half a gale, "You narrow-backed, no-chested, limber-kneed, parrot-toed, cotton-stomached, gimlet-eyed, blue-and-white-spotted sons of guns, what you backing and filling at us in that shape for?" The poor unfortunate fellows gazed at us in open-eyed wonder, and seemed not to know which way to turn or what to do. They wanted to run, no doubt of that, but there was no chance to retreat except around the end of the counter to the door, and we stood between. They no doubt thought these wild Americanos were bent on blood, and they would get as far from us as possible, so they ran back of the counter at the farther end, yelling "Carramba! Oh! Jesus Cristo, Santa Maria!" at the top of their voices.

On the floor was an iron cannon ball—for holding back the door, I think it must have been, as I could see no use for it other ways. Perhaps it weighed 15 or 20 pounds. This the crazy-headed 2d Mate put his eye on, and reached for it, saying to us, "See me make a ten strike"; and with both hands he started it rolling along the floor towards the end the poor affrighted clerks were at.

The ball went roaring along the floor like rumbling thunder. When halfway the length of the store, it curved towards the counter running fore and aft, struck it glancing, knocked two or three boards in with quite a crash and kept on its course. It had lost a little of its speed but not enough to stop it from sending most of the athwart-ship counter into a mass of ruins, amongst which the poor shrieking devils of clerks were piling over each other.

This had all happened so quickly that I had hardly time to take it all in until the final crash came. Then I did not know if it was best to run or not.

The author of all this mischief was dancing a breakdown and shouting, "Did you see them jump? Set them up again! Next time I'll pop some of them through the roof!"

This affair had straightened our 3d Mate up a bit. He told me not to let any of them go out the door, and started towards the wrecked counter and the poor gallied wretches. When they saw him coming, one of them attempted to spring by him but he caught him by the coat and made him stand still, which he did as well as he could for trembling. While holding him with one hand, he put the other hand into his own pocket and took out four or five dollars, making motions towards the others and the broken counter. This seemed to quiet the one he held somewhat, and he let him go, at the same time telling me to see he did not get into the street and call the soldiers, as there was plenty of that kind almost within hail. "All right," said I.

In a few minutes he had the lot quite tame, and as nothing could be said between that either could understand, by signs he gave them to understand we would pay for all damage done. It took about $15. to heal the wounds of men and counter, most of which our jolly 2d Mate paid. When he saw the fellow brought up by our 3d Mate, though, he started to strip, thinking there would be a fight. I soon quieted him down and told him to keep still. Much to my surprise, he did, so it was not long before peace reigned. The tenpin striker shook hands, and made signs for a drink all around.

One of the fellows went out and brought back ½ doz of fine wine, drove out some who had gathered in there, and shut the door. If I had not induced our officer to get away from there when we did, I believe these now good friends of theirs would have had them so tight they would not have known their own names.

After shaking hands and almost kissing each other half a dozen times, we started for the boats. The 3d Mate went first to the ship that the wild 2d Mate belonged to, and put him on board, then followed me on board our own ship. The Captain soon after came on board in a boat belonging to one of the merchants that lived a short distance out of the city. In taking his boat they had no chance of knowing anything of the trouble we had been in; and we did not say anything about it, you may rest assured. By this time the 3d Mate showed but little of the stuff he had stowed under his hatches.

The Captain told the Chief Mate to fill away the main-yard and keep the ship headed towards the end of the island. The yards being trimmed right, all sail set, ropes coiled up and decks swept off, the order came for supper. By the time we had finished that, we had got abreast of the point. The sun had been down a little time so it was now getting dark, and lights were gleaming from the shore line to high up on the slopes of land that extended back to the mountains.

As the ship was not more than a mile from shore when abreast of the point, the surf could be plainly seen dashing against the black volcanic rocks and sending the wild snow-white broken water masthead high. When the broad Atlantic wave, that had gathered for miles back its force, struck against an up-and-down cliff whose base stood in six or seven fathoms of water, and whose top was twenty-five or thirty feet above the sea level, you could imagine the sound it made, after spending its force; gigantic groans in despair of its death as it rolled back, leaving a deep hollow at the foot of the cliff, that would be almost instantly filled and replaced by another wave to meet the same doom.

When the point had got abaft the beam, much to our surprise came orders from the Captain to square in the yards, and the wheel to be put up. The men came aft to the braces and, when the yards were squared in, we saw that for some reason the ship was following the shore along. Our course did not lie this way.

After hugging along the land for an hour or more, the old ship making about ten knots, the order was given to hoist a light at the foremast head. A man was sent aloft; he lashed the light between the hounds of the foremast. This could only be seen shining ahead under the foot of the fore-topsail and foremast yard, as the hounds of the mast projected out on each side enough to cover it sideways; and of course it could not be seen through the mast. In a few minutes the Captain told the Mate to have a bright lookout for three lights set in a triangle. Shortly after, a man on the fore-yard sang out, "Light O!" "Where is that?" said the Mate. "Two points on the port bow," was the answer. The ship was headed for it.

Almost as soon as we headed for it, no more was seen of it. "Jump to that light on the masthead, and put it out! Don't be a minute about it, either," said the Mate. The man did as he was told and brought it down with him from aloft. The ship was now brought to the wind with the headyards aback, heading offshore, which was not over half a mile off. The 2d Mate, 3d Mate, 4th ditto and two of the boat-steerers were told by the Captain to go into the ship's run and take out fifteen or twenty boxes of tobacco and have them put into the forward cabin and covered up.

I was one of the boat-steerers that had been ordered to help with this work. We now knew what it meant. This tobacco was to be sold to smugglers from shore. It now was plain to me what made the Captain so indifferent as he had been during the day about being in any hurry to go on board. No doubt he had timed it with the merchants whom he had been with during the day, to have everything fixed for this event.

We had not got quite through with our work when we heard a thump alongside. Shortly afterwards the Captain came down the companionway into the cabin, accompanied by three finely dressed Portuguese, having every appearance of gentlemen. One of them seemed to do most the talking. His English seemed perfect. I was ordered by the Captain to stay and help him. The others went on deck. It did not take long for them to fix the business. I gave the number of pounds on each box, which they soon summed up and multiplied the amount of pounds by fifty; and the total of what this was, was paid in gold doubloons from a bag one pulled from under a large Spanish cloak he wore. They bought twenty boxes. Each box weighed 110 pounds—for which they paid nearly $900.00. This same tobacco cost in New Bedford about $260.00. Pretty fair profit [for the ship's owner].

The tobacco was passed into the boat that they had come off from shore in, and it did not seem much affected by the weight, as it was such a boat as the Portuguese use for deep-sea fishing. They shoved off quickly as the last box was lowered over the side. During the whole transaction no loud words had been spoken or noise made. All lights had been put out except the one in the cabin, and that had been concealed as much as possible. There was but little sound from the boat as they pulled for the shore in a different direction from that they came, and the oars, being muffled, were almost noiseless.

After laying to half an hour or so, the headyards were filled away, fore and main tacks boarded, and the ship was headed to the South away from the land. The wind had been dropping off since dark, and by the time the first watch was out it fell calm. When we came on deck (that is, the Starboard watch which I belonged to), it was flat calm. The sea was quite smooth and the stars were shining brightly, with hardly a cloud to be seen in any direction. Everything was so still that the ship seemed resting on a sea of glass, and only now and then would she roll half a streak and make dull bubbling sounds in the water alongside as she raised her side again, with the little pit-a-patter of the reef points against the topsails when now and then the sail slapped back from the raise and fall of the bows.

Everything seemed at rest. The land, no great distance off, had a mild look in the bluish-black velvety appearance that covered it; and I do not know if the officer and all did any caulking of the decks that morning watch before daylight broke. But I know I caulked the carpenter's bench the whole length of me until the sun commenced to show red in the East.

Men were sent to the masthead as soon as a white horse could be seen half a mile off. During the night the current had set the ship in shore, so the land was not more than five or six miles off. As the sun came glowing out of the clear calm horizon, a nice little breeze sprung up from N.W. The yards were trimmed to it, our course laid S.S.E. and by 12 M. we had a good fresh breeze two points on the quarter, with the old ship carrying quite a white feather in her teeth. Our course was for the Cape de Verde Islands, distance about 1000 or 1200 miles.

During the passage to this place we had not the luck to sight a whale. We had some light weather and some days we lowered the four boats for practice of the men at oars, so by the time the islands were reached, the crews of each boat could pull a good long whaleman's stroke: stern-two, pull ahead three, or stern-three, pull two; or stern-all, with good will—and all in shipshape order.

On the twelfth day after leaving St. Michael's we sighted the island of San Antonio whose steep peaks run almost straight skyward out of the ocean to the elevation of (I should think) eight or ten thousand feet, the shores of which must have been formed of what the hills could not retain for steepness. A more rugged broken cut-up land would be hard to find. Some of the peaks run into such fine points that it seemed a compass card might turn on them. Then again ridges could be seen apparently as thin on the edge as a saw and about as uneven. The wind did not permit us to reach the land that day near enough for the Captain to go on shore, so we lay off and on all night. Next morning we were close enough to lower the boat after breakfast and the Captain went on shore. He wanted to get some hogs and two or three more men, as these nigger Portuguese are very hardy and strong.

Just before dark the Captain came on board bringing in the boat three or four strong-looking islanders, black and with wool so curled that they could hardly touch their heels to the deck, some of the boys said. The only way they were different from the regular African Negro was in language. The boat also brought off some goat's milk, pumpkins, a few chickens and ducks. As soon as the boat was hoisted up to her place, the mainyard was filled away, all sail set, and our course laid South.

The weather continued fine and the gentle N.E. trade winds carried us quickly along. The watch on deck during the day were all busily employed, some fitting riggin', some making mats for putting on places in the riggin' and spars where any chafe might occur, some making spunyarn. The rattle of the machine used in making same afforded music for those employed.

One evening as we were about on the Equator, with our watch on deck, the 3d Mate, 4th Mate, the other boat-steerers and myself, standing on the weather side of the deck abreast the mainmast, some leaning against the braces coiled up on the pins in the fife rail, talking and smoking, the 4th Mate proposed having Old Neptune to come on board when we crossed the Line.

"Who is to take the part of Neptune?" said the 2d Mate. "Oh, for that matter, I will," said the 4th Mate. The 2d Mate said, "All right, I will see the Mate and Captain about it, and let you know what they say." The next night we had the middle watch. The 2d Mate told us that both Captain and Mate would help the fun along in any way they could. "I have got a rig most ready, and tomorrow can perfect it. If you say so, we will have it tomorrow night." "There are a number in the forecastle who have been to sea before. We must let them know about this," said the 2d Mate. "All right," said the 4th Mate. "How many are in this watch? Better have them come aft in a quiet way, and the boat-steerers can tell them."

Before the watch was out we had told the five or six able seamen what was up, and for them to tell the able seamen in the Larboard watch about it and be sure not to let the green hands into the secret.

The next day at noon, after taking the observation and working up the Latitude, the Captain said to the Mate in the hearing of the man at the wheel, who was one of the green hands, "By our reckoning we shall cross the Line by seven or eight P.M. I wonder if Old Neptune will come on board?" "Oh, yes," said the Mate. "He has got so now that he will not let any whalemen pass his Empire without stopping them." They knew that as soon as the wheel was relieved and the man went forward, he would carry the news of what had been said.

Just before the men came down from the masthead at sundown, the Captain hailed the boat-steerer who had the masthead at the main: "Do you see anything of the Line?" "I think I do," he replied. "Take the glass and see if you can make out anything like Old Neptune's boat." "There is something ahead, but too far off for me to make out what," was the reply. The Mate turned to the Captain, shaking his head knowingly, "That's him, no doubt."

As soon as it began to grow dark the green hands were sent below, about eight or nine of them. The forecastle scuttle was closed and guarded by three or four of the able seamen, the main hatch taken off, and one of the largest sized blubber tubs hoisted on deck. It would hold about sixteen or seventeen barrels of water. This was placed on deck just abaft the tryworks, about six feet below the top. The tub then was filled to the brim with salt water and two wide boards were run from its edge to about two feet above the after part of the works. A seat was made at this end, just high enough for one to sit on the ends of the boards. This seat was made to swing on its side next the projecting ends of the boards, the feet resting on the lower side or bottom. When this was lifted, a person could not help from tipping backwards and sliding down heels over head into the tank of water. Some steps were placed from the forehatch to the top of the works on the forward part of the tryworks, so one could ascend that way.

During the time these preparations were going on, the 4th Mate had put on his rig and come on deck. His feet were encased in two old mats made of spunyarn that had been partly worn out in the jaws of two topsail yards. They were lashed on with rope yarns and came above his ankles. Over his pants in front were two thrummed mats (made of strips of canvas with pieces of strands cut from unlaid rope, say, three inches in length, and sewed by the middle to the canvas just so far apart that when the two ends of each were unraveled they would meet: these were used in stopping the chafing of yards and riggin'). For a coat he had an old short oil jacket that had become quite dark from use. Over it and his shoulders hung down long pieces of spunyarn back and front, to represent seaweed. He had whiskers, reaching down to his waist, made of white Manila yarns, partly unlaid, sewn on a piece of cloth that tied over his face leaving only the eyes, nose and mouth exposed. His head had a wig of frowsy okum and short rope yarns in the way of hair, on top of which rested an immense Turk's-head that had been worked out of 12-thread rattlin stuff. In one hand he held a pair of grains [a four-pronged harpoon], in the other an old speaking trumpet.

All being ready, Neptune worked his way out on the bowsprit as far as the fore-topmast backstays. These went down through cleats each side of it, and the ends led beneath to the bows, where he took his place. As the two stays afforded him good standing and the use of his hands, in which he carried the trumpet (for he had left his trident until he could take his seat on the tryworks), raising the trumpet to his mouth, he bellowed out, "Ship ahoy!"

The Captain, who stood on the main hatch with his trumpet in hand, raised it to his lips and yelled back, "Hello!"

OLD NEPTUNE: What ship is that?
CAPTAIN: Ship *C. W. Morgan.*
NEPTUNE: Have you any subjects for me?
CAPTAIN: Yes, a few.
NEPTUNE: Haul aback your mainyard. I will come aboard.

The order was given to haul up the mainsail and haul aback the mainyard. The men shouted the order back at the top of their voices, and with a loud tramping and throwing down of ropes made such a confusion on deck that the poor devils below were about frightened out of their wits.

Neptune got in on deck and mounted the tryworks, sat down on a scrap tub turned bottom up, and sung out loud enough for them in the forecastle to hear, "Bring on the youngsters! I am in a hurry. Have lots of ships to visit tonight. One at a time."

None seemed willing to come first, but when told if they

kept Old Neptune waiting it would be harder for them, one ventured up the steps and had hardly struck the deck before he was blindfolded.

He was one of the Smart Aleck kind—should think, by his looks and actions, he had been one of the kind sometimes seen in country villages, swaggering into the country store, with pink necktie, scarlet vest, standup collar, cutaway coat and natty cane, with a damn-my-eyes cant to his hat, thinking every girl who should happen to look at him was dead gone, at least on his bold shape and corkscrew legs.

He came out of the forecastle scuttle with a swing-and-strut air of "Here I am! You cannot play your trick on me, if you do on the common green hands." He objected to having his eyes covered but it was of no use, for three or four of the old hands had hold of him by the arms and body. He had only a glimpse of Old Neptune in the light of the lantherns, which made him a bit quiet, so he was soon blindfolded and led up on the tryworks to the anxious seat.

Neptune spoke to him in a voice that sounded like coming through a cartload of rasps:

"Young man! There is every reason to suppose by your actions and talk since you came on board this ship, so I am informed, that you are a bad, bad man. But let us hope that when you have been shaved and christened, it will have the effect on you to cause a change in your former ways, make you a good sailor, and learn you how to be a man. I have a few questions to ask before we go on with our mild and soothing initiations. You are expected to answer promptly and open your mouth wide."

By the time Old Nep finished, the young chap had lost some of his bold swagger and began to think there might be more in it than he thought. It could be seen that he was getting nervous.

OLD NEPTUNE: What is your name?

GREEN HAND: Joseph Blake.

OLD NEPTUNE: Hereafter, while on board this ship, you will be called Joe, and to which may be added, "The Lady Killer," on state occasions. I have one more question to ask, and a little advice to give before we make a clean-shaved sailor of you. The advice is this: Never to eat brown bread when you can get white, unless you like it best. Never kiss the servant maid if you can the mistress, unless the mistress is not so pretty. Did you ever kiss a Negro girl? Answer loud!

GREEN HAND: Noooooo-oh-oh-oooo . . .

He seemed quite mad at the last question, and tried to answer loud, as ordered, by opening his mouth to its full extent; but before the "no" could well get out between his teeth, the tar brush covered thickly was rammed half-way down his throat. Gagging, spitting and struggling, he almost cleared himself of the boys who held him. His face was covered with a vile decoction composed of coal tar, slush and softsoap, and scraped with a piece of iron hoop, he groaning at the rough edges. During this time he had been asked by Neptune if he could swim. When he said he could, he was told, "On that perhaps your life will depend before you get done with this ceremony."

A bucket of water in the hands of one of the men, when the shaving was finished, was dashed full in his face, his heels were elevated and he went rolling down the incline, striking the water. He struck out like a man to swim, yelling for a rope, as the boys sung out a "Man overboard!"

Of course two or three strokes brought him to the tub's side, where he tore the bandage from his eyes and crawled out on deck. The wild look he gave Neptune, the tub of water, and the men shrieking with laughter, only made the mirth greater. He soon got over his fright, though, and was just as eager for the next one to show up as any.

The custom of having Neptune on board when crossing the Line is fast going out of date. No doubt, in some instances green hands were roughly used; but in no instance could they have been so dangerously made to suffer as in the old-time initiation of keel-hauling, as it was called, which has caused the loss of lives.

The manner of keel-hauling a victim the first time he crossed the Line was to take him out on the martingale guys—ropes leading from the bows, one off each side, to a spar hanging up and down, one end of the same being fast to the underside of the bowsprit nearly at its end.

A rope was made fast to his feet and another one made fast to his body, under his arms. The one to his feet was led aft outside the rail, clear of everything, so it would go under the ship. All being ready, the poor devil was tumbled overboard. He would sink far enough for the ship to pass her bows over him, and the rope aft, being hauled taut, would bring him square under the ship. The rope forward was slacked away on, just enough to go aft on the ship, as wished for.

As soon as the man came up to the rudder, a dozen or more men would clap on the rope and bowse him over the taffrail in on deck, most often insensible, by bringing him up heels first. A good deal of the water he had swallowed would run out. It was a barbarous custom and should have been stopped long before it was.

About a week after we crossed the Line, the welcome cry from the masthead came rolling down over the belly of the maintopsail:

"There! She blows! There!!! She blows!!!"

"Where away?" asked the Captain.

"Three points on the lee bow."

"How far off?"

"Three miles. There s-h-e b-l-o-w-s."

"Sperm whales?"

"Yes, sir, a large school. *There* she blows."

"Call all hands! Haul aback the mainyard! Get the lines in the boats!" were the orders given quick and loud by the Captain. There was but little need to give the order to call all hands, as all who had the watch below, hearing the welcome sound of "There she b-l-o-w-s," had sprung from their berths, hurried their clothes on, and scrambled on deck, in time to help haul aback the mainyard.

By the time the lines were put into the boats the whales had gone down. "Hold on, wait until the whales come up before you lower," said the Captain, who had his marine glasses fast around his neck and was going aloft up the fore riggin'.

Everyone was more or less excited, the green hands much more so than those who had taken whales before, and eager to catch sight of the low bushy spouts that would denote the appearance of the whales on the surface of the ocean again.

We were now in the Latitude of 16° South, Long. 34°35′ West. About this part of the South Atlantic a large amount of sperm oil has been taken, but of late years the whales have not been so plentiful, so to a certain extent it is not cruised over much. Whales, like cattle, which they resemble in some respects, seem to have favorite spots to feed in; and although they may be seen in any parts of the oceans between Lat. 50° N. to 50° South of the Equator, they seem to be at rest only when on their favorite grounds.

The Captain had been perched aloft in the fore-topgallant crosstrees with glass in hand for fifteen or twenty minutes, when suddenly he sang out, "There she blows! Four points on the lee bow, mile and a half off. Clear away the boats! The whales are still heading aft on the ship. Use great care, so as not to gally [frighten] them."

All was bustle and excitement. Every man sprang to his

boat. The merry rattle of the blocks, as the boats' falls passed through and each boat struck the water with a splash, made sweet music to the ears of the whalemen.

As soon as the boats got away from the ship's side and oars out, the sails of each boat were set and each man seated on the gunwale of the boat with paddle in hand, paddling with might and main to have his boat get to the whales first. Shortly after, we could see the low bushy spouts of fifteen or twenty whales right ahead, as the whales lazily blew them out, showing to us that they had seen nothing to alarm them, little thinking four greedy little monsters were pointing straight at them armed by twenty-four stalwart men who thirsted for their oily hides.

After we had covered two-thirds of the distance between the whales and the ship, the whales went down. The boats stopped paddling and sailed towards the spot where they last were seen, stopping with the sails shaking in the wind a short distance this side.

After fifteen or twenty minutes the whales broke water, some so near us we could hear them spout and see every motion of their massive frames that showed above water when they raised. The position of our boat was such that we were on their eye. A sperm whale can hardly be approached directly at right angles to his body, for they see better that way. A boat in that position is called "on the eye." The Mate's boat, however, being further to leeward and astern of us, was directly ahead of the school and in about the middle, so he took them head on. He hauled aft the sheet of his sail and had hardly got his boat under good headway when he was up to them.

The boat-steerer had his first iron in his hand ready to dart, when the whale's head had passed him far enough to reach his body with the iron. (It is almost impossible to dart an iron into a whale's head.) It was but a few seconds before up went his hands grasping firmly the iron, arms extended, body bent back, one foot firmly braced back against a cleat, right thigh set hard into a half-circle cut for that purpose in a two-inch pine plank fitted to the gunwale, about three feet back from the end of the boat, and called the "clumsy cleat." The iron was sent with force enough to drive the shank out of sight into the whale's body. Like a flash the second iron followed the first.

The whale in his pain threw his body half out of the water. With a terrific blow of his flukes he sent a volume of snowy water twenty feet into the air, then disappeared, taking the line so fast out of the boat that smoke arose from the loggerhead. After sounding until about half of the line, say, 100 fathoms, was run out, he rose to the surface, rolled, tumbled, ran his head out of water, snapped his jaws together like pistol reports, and dropped under water until his tail reared aloft ten or fifteen feet, thrashing the water into foam that spread over half an acre on the surface. As the other whales went off without our having a chance to strike again, we turned our attention to help kill the Mate's whale.

We soon had him fin out, and set the signal for the ship to come and take him alongside. The wind was fair so the ship ran down and we soon had him tied by the flukes, got up cutting tackle and commenced to cut him in. It took about four hours, as everything was new to most of the men. The green hands acquitted themselves well and were not afraid of the whale.

After we had the whale tried out and the oil stowed below, not having seen any more whales, we shaped our course for Tristan de Cunha. This island lies midway between the coast of South America and the Cape of Good Hope. It is about twenty miles in circumference and may have five or six hundred inhabitants, of mixed blood, white and Negro. We steered down towards the South on the wind, through the S.E. trades, not making any headway Eastern until we ran out beyond their influence. In Lat. 22° we struck a fresh westerly breeze and went booming to the E.S.E.

About ten days after leaving the place where we took our first whale, the island came in sight. It showed a rather barren prospect; not a large amount of anything except volcanic rocks, with here and there stunted trees and spots of grass. On the side where a cluster of houses stood, the land sloped from the hills to the shore; and here was about the only landing that could be safely made with boats. On the slope was the most pleasant looking grass land we had seen.

We stood in with the ship until the landing was about a mile off, and then wore around on the offshore tack, with the main-topsail to the mast. A boat was lowered and the Captain went on shore. He returned in three or four hours, bringing off in the boat some beef, a few pigs, ducks, chickens and some milk. These would help out our bill of fare while crossing the Indian Ocean.

The boat being hoisted up, all sail set, course E. by S., yards squared in, we went rolling along with a good brisk breeze from the S.W. Next day, after leaving Tristan, the watch on deck were employed in lashing any loose matter about decks and putting the boats on the upper cranes, to avoid their being washed away or stoven in our voyage across the stormy Indian Ocean.

The wind held quite steady for a day or two and the old ship staggered along with all sail set, except studding-sails, the braces as taut as bars; then the wind hauled out to West and commenced piping on. One sail after another was taken in until we had nothing set except a double-reefed fore- and main-topsail, and reefed foresail. The wind was howling through the riggin' but, as we were running square before it, the ship made easy work of it. Two men had the wheel and could keep her quite steady. It was only when running down the immense hills of water that would cause the rudder to gripe too hard, by the ship's getting a'sheer, that they would have to sweat to get the wheel aport or starboard quick enough to stop her broaching to and having the decks swept and masts carried away.

Our course was laid so as to sight the island of St. Paul, where, if weather would permit, the Captain intended to spend a day in fishing. The fish caught there are the finest kind. The island of Amsterdam, its only neighbor, is about sixty miles to the S.W. They lie in about Lat. 38° S., Long. 78° East, midway in a direct line between the Cape of Good Hope and Van Diemen's Land. St. Paul has a small harbor in which vessels of light tonnage can find shelter. Both are barren, and only the home of sea birds, seals, craw and other fish. They possess a burning soil and hot springs in which fish can be cooked. Shocks of earthquakes are of frequent occurrence.

On the morning of the fifteenth day since leaving Tristan, the island of St. Paul was in sight, about fifteen miles off, showing a rugged mass of dark gray volcanic rocks, with steep cliffs rising abruptly out of the sea, at the base of which the surf dashed in snowy foam masthead high, as the long unbroken waves that had been gathering power for hundreds of miles suddenly were met full force by these obstructions.

During the past twenty-four hours the wind had moderated to such an extent that all sail had been set on the ship, and the darkness from the heavy storm clouds that had been around us, hanging so low at times that it seemed our royal trucks would tear holes in them as the old ship rolled from side to side, had been swept away. The decks were dry for the first time in over two weeks. Clothes were brought out of the steaming forecastle and spread on the try-works and hung in various parts of the riggin', many of the men having had no dry clothes for a week past.

The bad weather that we had passed through made our

poor new-made sailors look forlorn and washed out. Those who had been to sea before found this nothing new. Still, it was hard, even for them. After standing their watch, often wet through as soon as they came out of the forecastle, they had no chance to change clothing, if they had dry to put on, until they were relieved and went below. There twenty-five men lived in quarters so small that it was impossible for all of them to find standing room at one time without standing on their sea chests as close as possible. And this was not for one day or month, but was their only home for four years, if the voyage should be that long (and it was forty-nine months).

As soon as breakfast was over, fish lines were rigged and a lunch prepared for men to go in two boats for fish. The ship ran around to leeward of the island and hove aback, heading off shore, about half a mile from the edge of the immense body of kelp that extends a short distance from shore and completely around the island. The beds of it in many places are several feet in thickness. This marine plant has a leathery look, shining like satin, with air cells, round and oval, on such parts as float on the surface of the water. It is impossible to pull a boat through the thick parts, on account of the oars catching their blades, but with a good breeze of wind a boat with the sail set can glide over as smoothly as a sled does over the snow; and here and there are places free, where one may pull through to the shore without trouble.

The two boats left the ship and soon were inside the kelp, catching fish as fast as the hooks could be dropped and hauled in again. After fishing until the boats were almost loaded to the thwarts, we set our sails and stood off shore. The ship stood in and took us up, all hands went to cleaning fish, and we packed down some six or eight barrels in salt. The rest were hung up to eat fresh. They were a treat—

though more to the green hands than to the older sailors.

The wind being still fair, the ship was put to her course E. by N. By midnight the wind had so increased that she had all she could stagger under. With a reef in the topsails, rolling from side to side and sending the foam from her bows in a roaring cataract forward beyond her dolphin striker, she now and then would dip the end as she pitched her bows deep in the hollow of the seas with stern raised at such an angle it seemed she would dive into the depths of the ocean; then settle down aft until the taffrail would be almost level with the water and raise her streaming bows, with bowsprit pointing skyward at an angle of 45°, tearing up the wall of water in vain endeavors to reach the top.

With the wind first on one quarter then on the other, blowing with such force for ten days that at times nothing was set on the ship except double-reefed fore- and main-topsails and reefed foresail, we had run out nearly our distance between St. Paul Island and the S.W. cape of Australia. The course of the ship was changed to E.S.E. After two days more running before the same gale, land was sighted on the port bow which proved to be the headlands of entrance to King George Sound.

Soon after rounding, the ship's head was brought up due North. As we drew in by the land with yards braced sharp up, we got the wind from off the land. We were within a mile or two of the shore, the sea was smooth as a mill pond, all sail was set on the ship; and a bright sunshine, drying up the water-soaked ship and men, with sight of green grass and trees, and the earthy smell, all combined to make every man on board feel life was worth living, after all.

We ran along the land nearly all day. At 4 P.M. we rounded an island into a deep bay surrounded with a snowy-white sand beach, almost too dazzling to the eye to look at in the sunlight. Steering over to the N.W. side of the bay, we dropped our anchor in ten fathoms of water, furled sails, got supper, and all hands turned in (except the anchor watch, which consisted of two foremast hands and a boat-steerer) for the first quiet night's rest since leaving the Equator.

Windward Chase

AT DAYLIGHT all hands were called to scrub decks and paint-work, as but little chance had offered to clean the ship in our rough passage across the Indian Ocean. From five to seven all hands were kept on the jump with soap and water, washing every part—the bulwarks, rails, houses and masts, as high as a man could reach up them by standing on the fife rails. As soon as the decks were swabbed up, all hands got breakfast.

After breakfast one boat was lowered, in which some shovels and buckets were put, and manned by six men, with an officer in charge who had visited this place before. They went on shore to clear out a spring of water and get it ready for taking on board some 100 or more barrels. The rest of the men were set at work breaking out the main hatchway, getting out empty water casks and putting beckets on them, to make up a raft to tow on shore to be filled with water.

The manner of rafting casks by whalemen is to drive a hoop off each end of a cask, the ones one-quarter way from chines to bung of the cask; having two pieces of small-sized rope, about a foot each in length, and placing the hoops over the ends of these ropes, which are doubled, the ends towards the bung, the bight towards the head of the cask, each directly opposite the other. Keep the ends two or three inches below the hoop, then drive the hoop down hard, turn the cask over and do the same with the other end. Bear in mind, both beckets on each side must be in line, so the raft rope can run straight through them. The casks are taken to the side, laid on their bilge, the raft rope rove through the beckets, one on each side, and then rolled overboard, one after the other, to the number needed.

By noon the boat returned, reporting the spring clear of mud and weeds; but the water would not be fit to put into casks until next day. The officer reported that on their arrival at the spring, and commencing to cut away the vegetation that had grown around it, a terrible crashing of brush and shrubs was heard, accompanied by loud hissing sounds, like escaping steam from a small engine. They all beat a hasty advance backwards, dropping shovels and hatchets. Some ran out from under their hats. After getting out on the beach, they saw an immense snake crawling slowly up a slight hill, now and then rearing his head four or five feet above ground and turning it from side to side, with his large mouth wide open. No doubt, said the officer, the snake was not more than twenty or thirty feet long, but it did look like it was a hundred and seemed in no great hurry to move on. They went back on tiptoes and worked for a while without touching their heels to the ground.

This bay is named Two Peoples, for what reason I know not, for not a soul lives here, nor is there a house to be seen. It is a half-circle in form and perfectly sheltered from the heavy westerly gales on the N.W. side. To the other end from where we lay was about five miles; then the coast made off sharply to the East. Thousands of cartloads of cuttle-fish shells, such as are given canary birds, are piled up along the beach at high-water mark and above, where they have been driven higher by southerly gales. These being crushed by heavy surf and ground to powder, mixing with the sand, formed the finest beach any of us had ever seen.

Some years ago whale ships came here in certain seasons to catch right whales, but it has been abandoned now for that purpose. The wandering bands of natives from inland used to come here in the whaling days and feast on the carcass of any whale that had drifted on shore, and would gorge themselves on it even if it smelt a mile a minute. So said those who had been whaling here. These natives no doubt are down to the keelson in the scale of humanity. There is nothing that has life in it but what they will eat, not excepting cockroaches, which they prefer roasted. They capture poisonous snakes by having a forked stick with pointed ends which they pin the snake's head to the ground with, and cut it off with a sharp piece of flint. I have been told they have not any feeling of modesty with each other, any more than barnyard fowls have. The power of endurance of these beggars, however, must be great. They will go without food quite a time, having a girdle around their middle which they haul taut as the stomach gets empty, so it is said.

Captain Sampson told us a story about one of them performing a long run in a short space of time for him, once when he was whaling in this bay. Having business with one of the leading men in a town some sixty miles away, he saw a tribe of natives on a part of the beach camping, so he went in his boat and landed, making signs so one of the number understood what was wanted and took a letter with him. His reward if he did the errand was to be a bucket of bread. Off he started. The Captain said he did not expect to see him for a week.

"Thirty hours after I gave the native the letter, a big smoke was seen at the beach abreast of the ship. Looking with a glass towards it, a native could be seen waving something in his hand. A boat was lowered and sent ashore. It returned with the answer to the letter I had sent." This man had been and returned 120 miles on foot in thirty hours. "A large bucket of bread (ship's biscuit) was given him, he sat down close to a stream of water on the beach above high-water mark, plain in sight of the ship, ate every bit clean, took a long drink of water, rolled over on his back face upwards and hardly moved for two days," the Captain said. He went on shore the next forenoon and looked at him, lying face up in the sun like a dead man, his belly puffed up like a bladder tightly blown up. How a person could hold such a strain and not fly to pieces was a wonder to him. It was a wonder the thing did not burst wide open, as hard-tack when moistened will swell at least twice its size. Maybe he did; for the Captain did not inform us any more on the subject.

We took on board about one hundred barrels of water, gave the ship a light painting on the outside, took up our anchor and stood to sea, shaping our course E.S.E. to clear Van Diemen's Land [Tasmania], bound to the cruising ground North of New Zealand. The wind still hung out to the West and with all sail set we rolled off one hundred and fifty to two hundred miles every twenty-four hours. In about a week we rounded the southern part of Van Diemen's Land and hauled up N.E. on a course to strike some islands off the North Cape of New Zealand, called the Three Kings.

These islets are nothing but barren rocks with sharp peaks, sticking out of the ocean from two to three hundred feet in elevation, due North of New Zealand, distance sixty or seventy miles. Around them has been a favorite cruising ground for whale ships in the past.

After reaching the cruising ground, we kept the mast-heads well manned, but for two weeks or more we looked for whales in vain. We had been lying to under storm sails for two days. A gale had been blowing from S.W. On the third day the wind began to moderate. By suppertime, 5 P.M., we had double-reefed topsails and foresail on the ship. A heavy swell was causing the ship to pitch and roll for all time.

We had just finished supper and lit our pipes, when the man at the masthead sung out:

"School of sperm whales on the lee beam, not half a mile off!"

Casting our eyes in that direction, we could see twenty or thirty good-sized whales tumbling about when the big seas would catch them and almost turn them over. Sometimes one could be seen on the crest of a wave. As it broke he would shoot down its side with such speed a streak of white could be seen in the wake he made through the water. When reaching the hollow between two seas he would lazily shove his spout holes above water and blow out his spout, as much as to say, "See how that is done." I have never seen whales at play before or since. It seemed too bad to interrupt their pastime, but they were the fish we had crossed three oceans into the fourth to find.

Heavy dark clouds and flying mist enclosed the ship so that nothing two or three miles away in any direction could be seen from her. Still, when the Captain gave the order, "Line in three larboard boats!" every man sprung to his station as though the sea was calm and the wind was light.

"Hoist and swing! Lower away."

Down went the three boats on the lee side, which was the larboard. It required quick and experienced movements to get the boats away from the ship's side without swamping or having them stoven.

"Use great caution," was the last word to us as we shoved off. The Captain sprang on top of the tryworks to observe our movements. As much could be seen from there as from aloft, the storm clouds were so close around us.

Our boat succeeded in getting out oars and pulling first.

It was no fit time to hoist sails, and when we crossed the ship's bows the other boats were two or three ship's lengths astern of us. So, if the whales did not go down, our chance to be amongst them first was good.

After pulling before the wind and sea a short time, I looked over my shoulder ahead of the boat, and as the boat rose I could see the whales tumbling and rolling no great distance off. By the way we were shooting over the water, not many minutes more would elapse before I should have passed through my trial, and be honored, or disgraced, as a boat-steerer; and, if failing, I would be disrated, sent forward, and never more have a chance to become above the common sailor on board a whale ship. These thoughts went through my mind, and although I did not fear a whale it made me nervous, as this would be the first time for me to strike one. The Captain was plainly to be seen on top of the tryworks with his spyglass, watching our boat as she approached the whales.

As we got on the top of a big sea, the 2d Mate sternly sang out to me:

"Stand up!"

Peaking my oar, jumping to my feet, grasping the first iron in my hands, mind made up to do or die, I saw three whales right ahead. I was looking down at them as they lay in the hollow of the sea, and could make out every part of their upper sides and plainly see their big flukes in motion as they slowly twisted them from side to side. They, like us, were heading to the leeward, and perfectly unaware of the sharp cruel iron that would soon penetrate one of their sides. I had hardly time to brace myself firmly against the clumsy cleat when the boat shot down the side of the sea, and amid the roar of breaking water with the boat's head a few feet clear of the whale, I darted first one iron and then the other chock to the hitches, just forward his hump.

Never in my life have I had such feelings of relief and

pleasure, as I saw the line run out when the whale dove into the depths, drawing it after him.

The Captain (so I was told afterwards by the boat-steerer of the 4th Mate's boat, which did not lower) had his glass on us just before we struck, and when I stood up he was all excitement, saying, "He stands up! Only strike that whale and I will give you anything I have, anything except my wife"; and as my irons struck the whale he threw off his hat, saying, "He is fast! Take my wife and all I have!"

I guess he forgot about his offers after we got on board, for I got nothing but the proud satisfaction that I had struck my first whale and proved that a boy only seventeen years old could fill a man's place on a whaleman's deck.

The whale sounded out about half the line, then came up to spout. By using great care with the boats in working around, none was stoven or capsized.

In an hour the whale was turned fin out. The ship ran down and took him alongside. It took us long after dark to have him securely fast and veered out from the ship to the windward. It was too rough to let him lie alongside, for the heavy surges the ship was making might part the fluke, and he would be lost.

During the night the wind went down, and also the sea. By ten A.M. we hauled him alongside and commenced to cut. At four P.M. we swung the last blanket piece over the side. He made about 85 barrels of oil. His jaw was seventeen feet long; his teeth would average in weight about two pounds each.

AFTER cruising around the Three Kings until the middle of January, we headed the ship on a course for the Bay of Islands, a port on the N.E. end of New Zealand. The Bay of Islands is a resort for whale ships to get recruits, wood and water, and give men liberty and paint ship. The entrance to the harbor is through a passage amongst small islets and rocks, for a distance of some ten or twelve miles. The scenery as the ship makes her way from the sea to the harbor is very pleasing to the eye, as you wind in and out, now by green-clad islets then by reefs of rocks, catching sight through the openings of the grass and woodlands on the main island, with here and there to be seen the little white cottages of the settlers, and cattle roaming at will around them.

These islands were discovered by Tasman in 1642. Captain Cook visited them in 1770 and made a survey. In the year 1814 English missionaries first went there, and got large grants of land allotted them by the chiefs. Some of it is held by their descendants now. How much good the missionary did the natives in early days would be hard to tell. What we saw of them did not show much in that line.

The Maoris are a warlike race, and the English have had some bloody fights with them, in many instances losing more men than the natives: but they are quite peaceful now. In some ways the natives resemble the Hawaiians more than any other natives of the South Seas. They have the taboo of the Hawaiians ("tapa," they pronounce it). The word Maori means the same: that is to say, anything to the manor born.

There are now and then found, on different parts of the islands, bones of an extinct bird called by the name of Moa; some of the leg bones are the size of a horse's, and they must have stood when alive ten or twelve feet in height. They could not fly clear of the ground, but it is said by tradition that they could run faster than a horse, and were fine to eat.

The old ship, under all sail and fine breeze, soon ran up to a high bluff of land we had been steering towards. Passing that, we opened up a deep bay in the land, nearly circular and five or six miles across, the land sloping to the shore almost around it. Two large rivers that emptied into the bay drained a large stretch of almost level land reaching for miles inland. Shortly after we opened out the bay, the yards were braced sharp up and the ship hauled to the West, pointing towards a sandy beach back of which rows of houses extended to the right and left for a mile or more, amongst which could be seen the usual gin mills that go with all English, and most American, settlements. We stood in towards the town until soundings gave us six fathoms of water, let go the anchor, gave forty-five fathoms' scope on the chain, furled the sails, and coiled all the riggin' neatly up.

Shortly after the ship was put in order, the Captain came on deck dressed up to go on shore to enter the ship at the custom house. A boat was lowered and brought under the gangway, the manropes put over the side, by which he descended, and the boat shoved off.

The Captain, who nad visited this port numbers of times before, was well known, so he stayed on shore until near suppertime. When he returned to the ship he was in a towering rage. During the time he was visiting his friends on shore, the boat-steerer who had charge of the men had gone with them into one of the public houses, and with the well-known hospitable feelings of the true English, the keeper of the house, his wife and others in there, had so filled them with provender out of a bottle that it was harder work for them to keep their oars in the water and pull the boat than it would have been pulling to windward in half a gale of wind chasing a forty-barrel bull whale.

We lay there some two weeks, took on board some three hundred barrels of water and a little wood, and painted the ship outside. Each watch had two-days' liberty and some spent most of the time, and all their money, in the public houses, and had roaring headaches the next day to show for it. There was quite a settlement of Maoris a short distance back from the English portion of the town. This is always added to in the season whale ships come here, so the residents say. Some of the men rambled about these, to most of them the first time of seeing Kanakas. Numbers of the women and girls were rather pretty, some of the half-breeds decidedly so, and most of them were quite well dressed and clean. There were many more females in the settlement than males.

FEBRUARY 10, 1850, we hove up our anchor and with a fine N.W. wind set all sail and stood out of the bay, rounded the bluff, and with a leading wind ran out through the channel we entered by, and soon had the broad Pacific before us.

Our course was shaped N.E. for the cruising ground around French Rock, a clump of ragged matter sticking up out of the ocean, looking at eight or ten miles' distance like a ship, and often taken for such. We cruised about there for a few weeks, but saw no whales; spoke some ships, they all reporting whales as being scarce; then shaped our course for the North to the Fijis, and to take in some islands in our way to get some hogs. Amongst ships we spoke around French Rock was the ship *Christopher Mitchell,* of Nantucket. She had been cruising on the coast of Chile and Peru. Not meeting with much success, she had come over on the New Zealand ground to try her luck this side of the ocean. In company with her we started towards the North.

The Captain of the *Mitchell* told us an incident that had occurred on his ship a short time before he left the Spanish Main to come here.

Amongst the green hands that sailed from home in his ship was a young man who had the appearance of refinement and deportment seldom seen in the forecastle of a whale ship, and although dressed in the loose attire of a sailor, looked the gentleman.

After recovering from a severe attack of seasickness he came on deck and soon was amongst the foremost attending to the duties of the ship, being about the first to steer a good trick at the wheel, and go aloft to loose or furl a royal. When reefing topsails he never seemed to heed how hard the wind was blowing or the sail flying over the yard. (When it is filled by the ship's head falling off from the wind, it tears back with force enough to hurl men from yard and foot ropes if they lose their heads or the grip of their hands on the jackstays.)

In the boat when after a whale he would pull his oar with the stoutest, showing no more fear of a whale than the bravest of the green hands, and in all things winning the respect of his fellow-shipmates and the good will of the Captain and officers. So much so that in one or two instances the officers had checked some of the boys from causing him embarrassment by joking on his feminine appearance.

Captain Sullivan told another instance of his courage that perhaps was more marked than any, so it seemed to my mind. One night when trying to double Cape Horn, a heavy snow squall struck the ship. At that time the ship was under whole topsails, fore and main course, and spanker. The ship was forced over so far on her side that her lee rail was under water and decks almost at right angles. The officer of the deck sung out to call all hands, let go topsail halyards. "Hard up the wheel!"

The wheel was put hard up, the officer and boat-steerers let go the main and mizzin topsail halyards, also the spanker sheet, but the men forward were too much frightened to do anything else but cling on the weather riggin'. As the officer yelled out again to let go the foretopsail halyards and haul up the mainsail, this young man, who had the watch below, came struggling out of the forecastle scuttle half-asleep, with but few clothes on, and heard the order. Catching hold of the fife rail around the foremast with his hands, and getting a brace with his feet on the combings of the mast, he sprung towards the starboard side of the tryworks, catching the cooler with both hands, hauled himself up the inclined deck far enough to get his feet braced firmly against the side of tryworks and cooler, and brought his hands within reach of the halyards and main tack, which he instantly cast off from their pins, both being fast near each other on that side.

The wheel being put hard up, the ship had fallen off some, but still hung until she was relieved by the fore-topsail running down and the maintack being let go, which was flying into ribbons with reports as loud as small cannons. She then quickly righted up and got before the wind. All sail was taken in while running before it except a close-reefed main topsail, main spencer and fore-topmast staysail, under which she was brought to the wind.

The Captain said, "No other man forward tried to do anything but hold on to save himself, except this one."

They had been out about eight months, so continued the Captain, when, cruising off Peru in Lat. 15° South, the weather was very hot. This young man had been sick for two or three days. One night in the middle watch everything was still as it could be on board a ship, with the sea calm and not a breath of wind stirring. The man at the wheel had nothing to do but lean over the spokes and keep himself awake. The officer stretched on the main rail with his arm over the monkey rail, now and then humming a few words of some song to show how wide awake he could be, when one of the men of the watch on deck, who had gone below to light his pipe at a light in the forecastle, came up the forecastle steps with a bound, rushed aft on a run to the place where the officer was sitting on the rail, stopped suddenly in front of him, and yelled in a voice loud enough for most of the men on deck to hear, "That young fellow who is sick is a woman!"

"What in hell is the matter with you?" said the officer, springing off the rail and standing in front of him ready to knock him down, he was so startled by the man in his movements and the words he had said. The man backed away a little and replied:

"He is a woman, sir. Come forward and you can see for yourself that he is."

Most all the men forward had come aft as far as the mainmast on hearing some of the conversation. The officer said to the man, "Come and show me what you mean."

Passing forward through the men (who seemed as much dumbfounded as the officer had been and were standing with open-mouth amazement) he quickly descended the steps into the forecastle and waited until the man had time to point his finger towards a berth that had its curtain partly drawn aside. He stepped quietly to the side of the berth.

Everything was quiet in the forecastle, except now and then could be heard one of the men who had the watch below tossing around in his sleep, uneasy from the heated quarters. Nothing had disturbed them, so none but those on deck were aware of what was about to take place. One of the lamps such as are used by the sailors to lie and read by shone into the berth that the officer stood looking into, and sure enough there lay before his astonished gaze a beautifully formed woman.

The officer told him, the Captain said, that for a minute or two he could hardly believe his eyes to think that before him lay a young unconscious woman who, in her feverish sleep, had disclosed what she had kept an inviolate secret, surrounded as she was with fifteen or twenty men for eight months. It was almost beyond belief but the facts were before him.

The officer removed the lamp from its place, cautiously closed the curtain, moved back a step and quietly called her by the name she had passed by.

At the unusual sound of his voice in the forecastle she awoke with a start. He placed his head close to hers and told her to hurry on enough clothes to appear on deck in, and go aft with him: her secret was known, this was no place for her. She hurried on her clothes without getting out of her berth and went on deck with him.

"They came into the forward cabin, the officer came to my stateroom and called me," said Captain Sullivan. "Astonished? Well, I should say I was," he replied to our Captain's question.

"What next did you do?" asked our Captain.

"We had a spare stateroom in the after cabin next to mine. I had that cleaned out at once and in a short time had a good bed made in the berth for her, and put the poor sick girl in it. Before I left her she put her arms around my neck and sobbing said, 'You have been good to me and I know you will hereafter.'

" 'Yes, I will, you are a sister of mine as long as on board this ship,' was my answer to the pleading one.

" 'Thank you,' she said, 'I don't think I could have stood my life much longer in the forecastle with those rough men, if I had not been taken sick. I am glad it has been found out and would have told you before, but was too ashamed and wanted to reach some port and get away without being found out.

" 'If you wish,' she said, 'I will tell you how I came on board this ship, as soon as I get better. What are you going to do with me?'

" 'It is about time for the ship to go into port for wood, water and recruits, and I shall immediately shape the ship's course for Lima, and put you in the hands of the American Consul, who will send you home to your friends. Now go to sleep, you are as safe as if you were at home with your mother.'

"I closed the stateroom door, went on deck too excited to turn in again, and discussed the event with the officers."

A number of things were mentioned about the young lady's actions that had caused remarks at the time they oc-

The *Charles W. Morgan*, rerigged as a bark and with other changes since Haley's voyage, remained an active whaler until 1921. She is shown here under sail, ca. 1900. (M.S.M. 56.1633)

"Evening Exercise": under the supervision of the mate, the boatsteerers wet the deck and the crew scrubs away the day's grime with their heavy brushes.

"Going on to a whale": under sail and paddles, with their oars peaked, a boat's crew quietly approaches an unsuspecting pod of sperm whales.

"Striking attitude": the boatsteerer (harpooner) prepares to drive his two-flued iron into a sperm whale's back

"Going over the whale's flukes": as the whale rises, nearly swamping the boat, the boatsteerer plants his second harpoon.

"Sports of Whalemen": a sperm whale, maddened by the harpoons in its side, attacks a twenty-five-foot whaleboat.

"Too late": as three boats' crews, all fast to the same whale, rest on their oars, the wounded whale dives before the mates can attack it with their long-shanked killing lances.

"Too late with the lance": a sperm whale crushes a whaleboat as the mate futilely drives his lance into the whale's head.

A mate churns his killing lance in a sperm whale's lung. Two other boats converge to ensure the kill.

"Cutting-in": the mates on the cutting stages cut free the continuous strip of blubber known as the blanket piece as sharks gorge themselves on the flesh.

"Heave pawl!": the crew at the windlass hoists the blanket piece aboard, peeling it from the carcass as the whale rotates in the water alongside the vessel.

"A case": the whalemen have hoisted the sperm whale's forehead on deck to bail the waxy liquid spermaceti from the resevoir or "case" within. Two men stand in the case and dip out the spermaceti into buckets for others to carry across the slippery deck.

"Trying out": with the tryworks flaming, the whalemen render oil from the blubber in the large cast-iron trypots. After cutting the blanket piece into smaller horse pieces, the processors have minced them to make "bibles"—pieces slashed vertically like book pages to expose more surface area to the heat—and the men at the pots render them, throwing the scraps into the furnace to stoke the fires. Alongside the tryworks are copper cooling tanks and a large cask for temporary storage of the oil.

"Extracting Ivory": whalemen remove the teeth and gums from a sperm whale's jaw. When cleaned, the teeth will be used for the whaleman's unique art form, scrimshaw, the art of carving or incising designs on whale teeth, bone, or baleen.

"Dog watch": during the second dog watch, 6:00–8:00 P.M., crews traditionally took their leisure on the forward part of the deck. These whalemen smoke and tell yarns around the windlass.

I asked the captain, "how far off he thought the Island were"? thinking a minute he replied, "about 45 miles, E.a. bearing about N.N.W. and I think plenty far enough off to afford us plenty of sea room, by our running to the N.E. thirty or forty miles, when the wind strikes us, which it will do shortly."

He had hardly finished talking when, the [illegible] of the maintopsail (it was close reefed) slapped hard aback against the mast, from a puff of wind that seemed to make the whole ship quiver from the force it struck the masts and riggin, at the time it struck us, the screeching from it through the riggin, was worse than forty tom cats sending up on a still night their musical strains in concert, it only lasted for a minute, and was gone again, so quiet was everything after it ceased, that the order from the captain to lay "square the mainyard," startled the men like a pistol shot fired close to a person unexpected. The mainyard had hardly been placed at right angles with the rails, when with a roar like a mad bull, the wind struck us square aft, with such force that the man at the wheel was pinned to it for an instant, as the house over his head had an opening in the back, through which the wind drew like through a funnel, righting himself he firmly grasped the spokes with both hands, sung out "Aye, Aye, Sir." to the captain who stood near him gave the order to "keep the ship dead before it."

When it first struck us the ships stern commenced to raise and her bows to settle.

Written in longhand on three different blank books, probably in the 1890s, Nelson Cole Haley's journal was donated by his daughter to Mystic Seaport Museum in 1944. This portion appears on page 84.

curred, but were sailorlike passed by. She had been seen to blush so often at rough speeches made in her hearing that it hardly had caused comment for some time. She was very reserved, but not offensively so. Some of the officers had noticed her aloft leaning over a yard in reefing or furling a sail with others each side of her. She seemed broader across the transom than those who were about her height. One day she was washing clothes in the same tub of water with another of the green hands. A boat-steerer standing by noticed in her rubbing clothes that she was doing it in shipshape manner, while the other was diggin' away with both hands. He (the boat-steerer) says to her, "By the way you wash those clothes, it would seem you had been used to that kind of work." He saw her turn red from her neck to the roots of her hair, but she only held her head down and made no reply. She was never seen afterwards washing her clothes but in as awkward manner as the greenest hand on board.

"How did she manage to hide her bust, or did she not have any?" said our Captain with a quizzical look on his face.

"We found in her things, when cleaning out the bunk she had occupied forward, a piece of topgallant duck that had been cut by her into shape to fit her form, lacing up on each side under the arm: and by drawing this tight she would flatten down her bust. With her undershirt over this and her shirt buttoned up, she did not show anything to denote her breasts. She told the Consul's wife after getting on shore that at times she suffered agony with the heat and tightness of the bandage around her and had to take it off."

"How did she dress after coming aft?" asked our Captain.

"I gave her some of the calico and white cotton we had on board for trading, and she made herself quite an outfit, and when dressed in her proper attire she was a pretty girl, even if her face was rough and sunburnt. Of course her hands were rough, but they still were a pretty shape."

"How long was she on board after you found her out?"

"About two weeks."

"Did she tell you what had set her adrift on such a mad and almost unheard-of freak?"

"Yes, she told me a part of her life, after she had recovered enough to be around. She had belonged in some town on the Hudson River; her father had been quite rich but had died some six months before she left home. Her mother was alive and was quite well off. She had a finished education and, by what she said, must have been a hard woman, but no doubt a wise one, as she had forbidden the house to a young man who had been keeping her daughter company. It seems the girl thought too much of him to give him up and she left home with him, he promising to marry her when they got to New York City. He left her in a few weeks and never married her. She had a little money she had brought from home, but wanted to find the cuss who had deserted her and make him keep his promise. She hired a detective and had him traced to a shipping office in the city that shipped green hands for the whale ships and sent them to the various ports that whalers fitten from. Her mind was then made up to follow him.

"Being quite tall for a woman and built slim, she put on men's clothes, went to the shipping office, and passed herself off as a green boy who wanted to go a-whaling." (There would be but little trouble for anyone to ship in those offices, as so much a head is given them for each man sent to the outfitters in such ports, but she made a terrible mistake in thinking she could find him this way, as there are at least eight or ten different whaling ports, and five or six hundred ships; some sailing to all parts of the known oceans.) "She was sent to New Bedford, from there to Nantucket, and hardly had time to get her half-crazed head clear before she was at sea."

"Did she say what she would do if she should find him now?" asked Captain Sampson.

"I asked that question," was the reply. "For a minute she held her head down, then looking up with glitter in her eyes she quietly said, 'I would kill him like I would a venomous snake!' And I think there was no question but what she meant it," Captain Sullivan said. "I would not be in that man's shoes for money if she ever does run afoul of him."

Our Captain said he thought that young man the girl was after would be safe only behind prison walls, anywhere near her.

"How did you dispose of her after your arrival in port?"

"As soon as possible after we dropped anchor and furled sails, I took the ship's papers on shore, entered my ship, and talked over with the Consul (and his wife, whom he had called in) what had best be done in regards of the young lady. The Consul's wife said, 'Have her brought here to stay with me, until you gentlemen make other plans for her.'

"Returning on board, I told her what had been done, and she burst into tears, and between her sobs thanked me over and over again for the kindness I had shown her. I told her what I had done was my duty to do: any true sailor would have acted the same, and if she wanted to please me, to dry her eyes and put on the rig of a sailor for the last time and go on shore with me. As soon as she was dressed, she was to send the boy to let me know—I was going on deck. 'You have nothing to wear on shore that is decent, but that kind of rig, is my reason for asking you to dress in it.'

" 'I will do as you request, but hope I never shall wear any but my own clothes after today,' she replied.

"In a short time the boy came on deck and told me she had sent for me. Upon going into the cabin I was surprised to see how pretty she looked. Her cheeks had bright spots of red in them that showed in fine color with her white skin, that had lost a great deal of its tan during her sickness and keeping under cover from the sun. Her eyes were black, and shone with such flashing brilliance they were startling. On her head she had a fine braided Timor sennet hat, with a black ribbon an inch in width, the two ends hanging two or three inches over the rim from a double bowknot fastened behind it. The shirt one of the boat-steerers furnished her was white with a wide collar of blue hanging well over the shoulders and down the back, under which and around the neck was a black silk handkerchief tied in a sailor knot, the ends hanging halfway to the waistband of her pants. The bosom of the shirt was also of blue, with cuffs of the same color turned back on her arms some three or four inches.

"The pants were of black broadcloth and belonged to the same boat-steerer who had donated the shirt. He was about her height, but not quite so broad as she was in the hips, which made them fit her form without a wrinkle. From halfway down her thighs, above which no wrinkles showed, they widened to where they reached the feet, so that when she pulled forward the bottoms would just overlap the toe. With a pair of white socks and low-quartered shoes on her feet, she looked, as she came towards me smiling, too sweet for anything and I had to do what I had never attempted before, catch her in my arms and give her two or three good Down-east kisses and ask her pardon in the next minute, which she not only granted but reached up and kissed me twice, saying, 'God bless you who have been more than a brother to me, a poor ruined girl.'

"When we went on deck the men came aft and one by one shook her by the hand. She held down her head most of the time but bid them each a low good-bye. Each officer shook her hand and bid her Godspeed. When she bid good-bye to the officer that brought her aft, she clung to his hand for a few moments and said in such a low sweet voice that all in hearing were charmed:

" 'The officers of this ship are Gentlemen, and have my heartfelt thanks, but you in your kind actions on the night of my discovery, I shall remember as long as I am alive!' Letting his hand go, she went down the side into the boat and we started for the shore.

"Before we had got more than halfway from the ship to shore, a large number of people could be seen around the landing. The news had spread that a woman was to come on shore that had passed eight months in that ship as a man, and it created such excitement that men, women and children were flocking to see us land.

"When we got out of the boat, the crowd was immense but they opened out right and left, leaving passage enough for us to pass through, closed in behind us and followed us until we arrived to the Consul's house, and crowded thickly around it. The Consul's wife took the girl in charge and when she appeared at the table for lunch (which on the invitation of the Consul and his wife, I took with them) dressed in some rich attire, it did not seem possible that such a change could take place in the manner and looks of the same person in changing dress. The hair cut short and sunburnt hands were there, but nothing else to prove this person to be any other than an accomplished young lady who moved in good Society.

"We had a long talk together of things that had occurred while she was on board the ship. Afterwards the Consul's wife and she retired, leaving him and me to discuss in regards the next step to take in her affairs. I told the Consul her share of the oil taken would amount to about one hundred and forty or one hundred and fifty dollars; and take from that the advance paid for her outfit to the land shark who shipped her, it would only leave her about fifty or sixty dollars. This money I would pay to her at once.

"He says, 'All right, I am empowered to send all distressed seamen belonging to our country home and she comes under that head, sure; and I think it is a good plan to get up a subscription paper, which you and I can head with the amount we see fit, and send it around to the merchants and any others who see fit to contribute.' This being done, some five hundred dollars was collected. The officers and men of my ship all gave something; so did most of the captains, officers and men of the other ships that were lying in port at the time. Before we sailed, a chance for her to take passage to Panama offered and she sailed for that place. Arriving there, she could cross the Isthmus to Aspinwall, there take a steamer for New York.

"Before she left for home I gave her my address. I saw her on board the vessel that she took passage on. Leaving the wharf to go on board with her, when the boat had just shoved clear of the wharf on which a big crowd stood, three rousing cheers were given by them. She answered by waving her handkerchief."

"Well," said our Captain, "I have been to sea, man and boy, for over forty years, but never have seen or heard tell of the like of that, and by ye gods if you had not told it from your knowledge, I would not have believed it possible. One thing sure, there is not many girls could do it." *

WITH a fine breeze on our quarter and all sail set, we soon ran up into the Lat. 20° South, Long. about 175° West from Greenwich on the morning of April 10th. The wind was just enough to make the sails swell out from the yards like a well-formed woman's breasts, and almost as graceful. The sea was smooth; now and then the reef points in the topsails could be heard to patter on the sails as the ship rose and fell from the ocean swell. The sky was deep blue with white-looking clouds scattered over it here and there, seeming motionless. All hands had a pleasant word for each other and faces bright and cheerful, as this was the brightest day on the sea we had seen for weeks.

The decks had been scrubbed and washed down, the

* Newspapers in the States carried a report of this incident in 1852, at about the time the *Mitchell* returned to Nantucket.—*Ed.*

brooms and swabs put away in their places, and the men were lounging around the windlass cracking jokes and spinning yarns, waiting for six bells (7 o'clock) to strike, and have breakfast. The boat-steerers were sitting in the head of each boat, oiling their irons and lances, now and then taking off the wooden sheath and with a small stone whetting one, now and again. (One of the boat-steerer's duties is to have all his irons and lances kept sharp and the shanks scoured, until they shine with brightness. His first duty every morning is to attend his boat after the decks are washed off.) The Captain and Chief Mate were walking the quarter-deck, talking to each other. The 2d and 3d Mates—the 4th Mate being aloft, on the lookout for whales—were talking in the waist, the cook, steward and boy dishing up breakfast, when the peace and quietness of the deck was thrown into the wildest confusion and excitement by the thrilling and welcome sound of the 4th Mate's voice at the main:

"T-h-e-r-e s-h-e b-l-o-w-s! T-h-e-r-e s-h-e b-l-o-w-s!"

"Where away?" said the Captain, jumping up and down in his excitement.

"Two points on the lee bow, sir!"

"There she B-l-o-w-s! There she B-l-o-w-s!"

"How far off? Sperm whales, by the gods," said the Captain, now wild with pleasure.

"Two miles and a half," said the 4th Mate.

"Haul up the mainsail! Haul aback the mainyard! Get the lines in the boats!" said the Captain in rapid succession. Then singing out to the cook and steward, he told them to have breakfast ready right away. During the time his orders were being carried out, the 4th Mate had come down from the masthead, also the boat-steerer; the shipkeeper had taken the lookout, and reported the whales had gone down.

By the time we were through breakfast, the shipkeeper reported the whales up and they could be seen from the rail, four points on the lee bow, a large school, heading partly to the leeward about a mile and a half from the ship; and by the way they shot their low bushy spouts just clear of the water, they showed they were perfectly still—in other words, were not frightened.

The Captain took a look at them, and gave the order to his officers: "Lower away the boats!"

The men being all in their places, each crew on the ship's rail abreast the boat they belonged to, the officer of each took his place in the stern and a boat-steerer in the forward end, both with hands to the blocks to unhook the instant the boat slacked the falls by striking the water. By the time the tackles were unhooked and out of the way, the men had scrambled down the ship's side into the boats. In the space of five minutes all boats were clear of the ship, oars out and peaked.

As the whales were to the leeward, our sails were set on the boats, and with men sitting on the gunwales with paddles in hand, we fairly made the boats fly over the water, each boat trying to get ahead of the other and be the first to strike a whale. (Paddles are used to approach whales in light weather. Each boat has five, and when not in use they are tied under the thwarts close up on each side, out of the way of everything. One can approach a whale with them, properly used, when the noise that oars make in pulling would be heard. The hearing of whales is keen—more so when they are under water a short distance from the surface.)

We made such good progress through the water that their spouts could shortly be seen from where we sat on the gunwales, and it did not take long to near them enough to see their humps sticking out of the water, which they hardly buried beneath the surface as they just moved along, heading directly away from us. The position we occupied was the best manner of approach for striking. There was hardly any difference in the line of the four boats' heads; and it did seem the school we were going to were ranged so near parallel to each other that the chance was good for all four boats to strike whales together, as there were fifteen or twenty whales ahead of us, and the boats kept plenty of distance from each other, so as not to be in each other's way.

Everything was still, nothing to be heard but the sound of the noise made by the sharp bows of the boat cutting through the water and the whales spouting. All looked for fine striking when, just as I was expecting the 2d Mate to call me to my feet, to make ready for striking the whale not over a ship's length ahead of us, one of the men in the Mate's boat struck the bottom of the boat with the edge of his paddle. Before the sound had hardly ended, every whale disappeared like so many grindstones dropped into the water, leaving nothing to show for the large school of whales, that a few minutes before were scattered over the water for acres, but a lot of foam and bubbles here and there on the surface.

Everything in the boats was quiet for a short time; then the air commenced to turn blue. It got so blue shortly that some said they could smell sulphur. The poor devil who had accidentally hit the side of the boat with his paddle, and so gallied the whales, wished himself behind an altar in some church, to escape the oaths and curses showered on him.

After we had lain to with the sheets of the sails slacked off for half an hour or so, the ship ran up the signal that the whales were in sight on her weather beam. Down came our sails; and with oars pulling, the boats were heading to the windward in chase of the startled whales.

None but whalemen can understand what the feeling of chasing gallied whales to the windward means. They know it means long hard pulling, and often nothing for it. Sperm whales when frightened (or as whalemen say, "gallied") as a rule run to the windward. I have seen them gallied in a calm and turn in the direction from which the wind blew last, and keep that course for hours. It has puzzled many men as well as myself to account for their being able to keep a straight course for hours and never vary a point by the compass from it.

As the ship was lying in our course, we pulled within hail of her. The Captain sung out to us, "The whales are about three miles off." He did not think we had much chance to catch them as they were going quite quick, but added, "Go ahead and try them."

With set faces we straightened out on our oars, not a smile on any face, you may believe. About 10 A.M. the wind died out to almost a flat calm, the sun poured down its heat like a furnace, the perspiration was streaming down my body and dropping from my heels. The others in the boat pulling were not any cooler than I. We all suffered alike. We had got into better humor now, as we had so far outrowed the three other boats that they were all behind. The 4th Mate's boat was next to us. He was at least a quarter of a mile astern.

About 12 M., with the ship just showing the heads of her topsails and topgallant sails above them on the dark line of the horizon, the 2nd Mate looked steady ahead for a few moments and quite excited sung out, "There she blows! Not more than a mile off. We are coming up with them hand over fist. They are getting still. Pull hard, my hearties! We will have a whale yet."

This news put new life into us and we put more strength on our oars. The boat moved with quite a speed from what it had been doing for two hours past, judging by the way we dropped the boat nearest us, which no doubt had not yet caught sight of the whales and was pulling for a forlorn hope.

In about half an hour the 2d Mate told me to look over my shoulder ahead. Doing so, I was wild in a minute, for within a short half-mile could be seen the school of whales. Looking again in a few minutes, I could see we had shortened the distance quite a bit between us. I spoke to the crew: "Wake her up, boys, one more spurt and we will be alongside of them."

They put forth all the strength left them and for a few minutes the old boat fairly flew over the glassy sea. Looking shortly over my shoulder, I could not see a whale. Speaking to the 2d Mate, I said, "They must have gone down."

"Yes," he said, "They have been down about five minutes."

After pulling a few minutes longer, I spoke to him, saying, "By the way we were drawing on those whales, if we keep on pulling will there not be a chance of our passing over them while they are down?"

He replied, "Maybe there is, but I do not want to stop pulling and have those other boats come up and perhaps take from us our chance of striking a whale first."

I then said to him, "The nearest boat to us cannot get here before those whales will be up again, and I believe if we keep on we will run over them and start them going again. The ship is most out of sight and this will be our only chance."

He got mad and spoke quite sharp: "Stop your growling! I will peak the oars and take the paddles."

It was some relief to change our position from the oars that we had been so long pulling. After we had been paddling a little time I got uneasy, and wanted him to let the boat lie still until the whales should come up again.

By this he got furious and wanted to know, "What in hell do you mean, trying to take charge of the boat? I am your boss, so shut up your head or I will come forward and tumble you head over heels overboard."

Just then, looking over the side of the boat, I saw the shape of a whale quite deep down, but saw it plain enough to know we had got into the position I was afraid we might by keeping on. Turning to him, I told him I did not know as he would come forward and throw me overboard, but I did know that I be d——d if I would paddle any farther over whales under water and thereby frighten them. Springing into the head of the boat, I caught up the first iron in my hands, in case a whale might come to the surface within reach, looked over into the water, and saw three or four whales on their sides looking up at us.

The next instant I saw one which seemed to make up his mind that he would eat us, and in that way stop our chasing them any more. He was about twenty feet below the bottom of the boat and heading directly for it, his mouth wide open. As the jaw was white, it could be seen with its two rows of teeth, and looked mightily wicked.

Knowing we were going to catch it in a few minutes, and holding the iron in my hand in hopes I might be able to dart it into one of the whales that I could see under the water ahead of the boat before the whale under the boat stove us, I said, without turning my head to the 2d Mate, "Look out for that whale coming up under the boat with his jaw open!"

The whale I was looking at in hopes to strike was too far off for much chance to strike him, but I knew it was about time that the one coming for us under the boat would hit us. I darted the iron towards the one ahead as hard as I was able, hoping it might accidentally stick into some part of his body, but he was too far under water and ahead of the boat. I had not time to straighten myself up from the force given the iron when the crash came, and a funny thing happened to me. I was shot upwards at such an angle that I turned completely over in the air and came down headfirst into the water, amongst the pieces of the stoven boat, striking my hand on a piece of it with such force that it was two or three weeks before I could use it to any great extent.

The whale had hit us square under the keel with the end of his head, just clear of catching the boat inside his mouth, which no doubt was lucky for some of us, for if he had done so and brought his jaws together, there would have been crushed men or broken limbs at least.

By the time I came up from my bath and got my eyes clear of salt water, the whale had done his work and left. Swimming to the biggest part of the wrecked boat where the 2d Mate and men were holding on to keep their heads above water, I took a hold myself. We were a sad-looking lot of men. The sea around us was covered with oars, buckets, water kegs, paddles and broken pieces of cedar, some of the lance poles sticking out of the water, the boat keel smashed in; and part of one gunwale and side were torn out, where the jaw had caught when he gave the parting snap to her before he left.

I took a look at the 2d Mate's face. He looked downhearted enough. My anger had cooled from the bath I had taken, I think, and I felt sorry for him.

"Well, sir," I said, "this is too bad after all the hard luck we have had today."

He said nothing for a few minutes, then turning his head towards me, said, "Nelt" (my nickname on ship that voyage), "I was a d——d fool from awayback, what I said to you before this happened off. You was nearer right than I."

My reply to him was, "I am sorry for all that has took place today, but most so on the loss of our fine bully boat."

"That's so," he said. "She was a good boat but it is all right, we can get another boat. But if that beggar of a whale had took some of us, it would have been a good deal worse."

Shortly after we had got pretty well soaked, the 4th Mate pulled up and took us in his boat. When the other two boats came up, two of the men got into each of them. We picked up the oars, paddles, sail, buckets, etc., floating around, took out the spare irons and lances from the stoven boat and left the wreck floating. She was not worth towing to the ship, which could be just seen like a white feather on the line of the horizon to leeward.

For an hour or more a breeze had sprung up from the same quarter it had been in the morning, so with our sail set and pulling we reached the ship about 10 P.M. She had hoisted signal lantherns as soon as it was dark, and was lying with her mainyard aback. After hoisting up the three boats all hands got supper and the watch went below, all hands tired and cross. The Captain was more cross than any; he came after me red hot, and claimed that I must be no good, to let a whale get near enough to a boat to have him stove it, and I not get an iron into him. I said nothing more than that, if I could get a chance to strike a whale, I did not fear but it would be done. He growled some more and went below.

The next day a new boat was taken down from the overhead (as it is called on board whale ships, where two or sometimes three boats are carried when leaving home, to replace boats lost on the voyage) and refitted for us. There are many little things needed in a boat, and absolutely necessary in catching whales, that a person not knowing about would hardly notice in looking at a boat properly fitted. So it was some days before my boat was in as good order as the one that had been lost. I do not know if the 2d Mate ever told the Captain how the boat got stove or not, but I never even told the other boat-steerers; and if the boat's crew did not tell of it, I think that the facts were never known.

Under the Lee of the Devil

THE DAY after leaving the cruising ground where we lost our boat, land was seen about 4 P.M. from the masthead, and so reported. By sunset the Island of Eoa, the land sighted first, bore N. by W., distance 20 miles, and Tongatabu, seen later, bore N.W. by N., distance 25 miles.

The island of Tongatabu, like Eoa, belongs to a group called the Friendly Islands. Included in this group are a hundred or more islands, some quite large; those seem to be of volcanic formation. (One of the group has a volcano and we saw the fire some forty miles at sea one night.) A number are not inhabited. The old Dutchman, Tasman, discovered them, and Captain Cook named them when he visited them in 1797. About the same time the missionaries went to the Hawaiian Islands, in 1820 or thereabout, the English located on Tongatabu and, I have been told, ousted some Catholic priests from there. The natives, however, took kindly to religion, or part of them did with a powerful chief at the head: and with Bible in one hand, sword in the other, soon converted the others, or laid them with their fathers. So for years this same kind of propagation has been going on, until the entire population have fallen into the ranks; and sailors, as a rule, when the ships visit there, do not care much for a day's liberty on shore. The inhabitants are very kind and of course friendly, but look on all outside the church as devils, and have no scruples in telling you so, either.

At dark we took in light sails, furled the mainsail and hauled the ship on the wind, heading off shore, to wait until morning for landing on the island. For the past day or two the weather had changed and the sky had lost its clear bright blue, which had been replaced with a reddish mist somewhat like fog but without dampness, through which the sun shone with a red color as if shining through air filled with smoke, casting shadows on the deck from riggin' and sails that had a weird look. Although the wind was light, now and then a faint moaning sound could be heard in the riggin' from it. Since 12 M. the weather had cleared and the Captain told the officers while eating supper that he hoped the weather would be better now, and that the hurricane he had feared from the looks of the weather in the last day or two would not strike us.

After six bells the Captain came on deck. The sky had shut down thick and dark. Not a star could be seen. The air seemed oppressive and not free to breathe. He walked the deck a few times, then gave the 3d Mate orders (as it was his watch on deck) to doublereef the topsails, furl the foresail, jib and spanker. By nine o'clock all had been done as he ordered, the braces hauled taut and ropes coiled up. The Captain walked the deck for half an hour more. Then, starting to go below, he told the 3d Mate to hoist all the boats up on the upper cranes (this is only done when heavy gales are expected), closereef the main-topsail, furl the fore and mizzin topsails, and see everything loose around decks firmly lashed. "If I miss not the looks of the weather," he said, "we shall catch the hurricane before morning, after all. Let me know if any change takes place in the weather." Then he went below.

Before our watch was out, at six bells (11 P.M.) everything was put into good shape to meet the expected storm. Spare gaskets were put on all the sails and hauled taut, after rolling up bunts and yardarms of same as close to the yards as possible. When we went below, our watch being out, the only change in the weather was, the swell seemed more heavy and the air closer. The wind had not increased.

On going below, the steerage seemed so hot and close, after turning in, that it was hard work to get to sleep. At three o'clock, when our watch was called, on turning out, the ship had such an uneasy motion to her that you stumbled about in putting on your clothes. I said to the boat-steerer who had come below to call us, "The ship has a very uneasy motion. What is she doing, trying to stand on her head and roll over, keel out, at the same time? What kind of weather is it?"

He said in reply, "There is not wind enough to blow a rag out of the maintop, but of all the nasty sea that you ever see at sea, when you go on deck, you will see the d——dest sea, at sea, that a sea can be."

"See here!" said MacCoy, the other boat-steerer that had the watch below with me, at the same time catching up a boot by the top and drawing it back over his shoulder as if to strike. "If you ever come into this steerage in the dead hours of night, where innocence and virtue abound, to disturb and violate the holy thoughts that we have, on turning out to go on deck on a stormy night like this, I say, if you attempt again, in a cold-blooded manner, to recite such a heartrending sea of the tale, or tail of the sea, your life will not be worth a minute's purchase. We will see that the boot makes you see more stars at sea than you ever did see since you have been at sea. You see that door, now see how soon you let us see you go out of it, where you may see all you want to see about the sea. Your life is in danger, do you know it?"

He went on deck laughing, leaving us to put on our clothes and recover from his base midnight attack, the villain.

When we reached the deck, the darkness was intense. The ship seemed in a world of her own. Heavy black clouds hung above us so low that the skysail poles seemed to penetrate them. Around us nothing could be seen, half a dozen ship's lengths away from the ship in any direction, except the dark waves whose tops now and then would break with a spiteful snap, showing phosphorescent bright and ceasing almost instantly.

The ship was rolling fearfully, both to windward and leeward, as the little wind that was blowing from S.E., the same point that it had been from all the day before, was not strong enough to steady us under the low sail we had out. Now and then the ship would pitch her head down, almost burying her jibboom under water, and raise it again with water streaming from her bows in torrents, the guys and stays dripping from being submerged when the dolphin striker had been sent nearly its length under water in the fearful plunge forward the ship had made. Then her stern would settle with a force and send the water flying feet away on each quarter as her counters settled almost far enough below the surface for the water to come over the taffrail. It seemed something must give way before the terrible force of the sea, which at times made the ship tremble from stem to stern; and when her masts and yards performed circles and all kinds of angles in the air, she seemed like a distressed person tossing her arms for help.

With the yards and blocks creaking, ropes slatting, shrouds and backstays weaving, rudder jumping and groaning, wheel ropes slacking and tauting from strain brought on them by force of the waves as the ship tumbled about, the man at wheel held on to it with a firm grip. If by inattention or carelessness he should lose his hold of it, most likely it would fly around and break his arms or legs before he could get out of the way; and there might be other damage at such times, besides. All in all, it was a dreary deck we had come on; and what was coming, we knew not. All were well aware that something out of the common was about to take place.

The Captain was on deck. Now and then his face could be seen, from the light in the binnacle when he looked at the compass. His face looked worried but showed no fear. I hardly think, from what I knew of him all the time I sailed with him, that he knew what fear was. About eight bells (4 A.M.), a light streak was seen in the dark mass of inky black clouds to the S.W. It had hardly appeared before the Captain yelled out:

"Hard up the wheel! Let go the starboard braces! Man the port main brace! Haul up the main spencer!"

The wheel being put hard up, the mainyard shivered in, the ship slowly swung off from the wind. When the light spot was directly astern the wheel was steadied, the yards trimmed with the wind about two points abaft the starboard beam. The Captain told the 2d Mate to have the men stand by the main braces. "Never mind coiling the ropes," he said. "We will soon have on us a hurricane that will try men's souls, and our good ship likewise. It is coming no doubt out from where that bright spot shows, and I want to take the first of it astern, and run a bit to the North to give us more sea room."

The 2d Mate, after he told the men to stand by the braces and keep their eyes open in what they were told to do, walked across the quarter-deck to where the Captain was standing on the mainrail, just forward the mizzin riggin'. Holding on the forward shroud and looking astern at the bright streak that was growing dim, he asked the Captain, "How far off do you think the islands are?"

Thinking a minute, he replied, "About forty-five miles, Eoa bearing about N.N.W. and I think far enough off to afford us plenty of sea room, by our running to the N.E. thirty or forty miles when the wind strikes us, which it will do shortly."

He had hardly finished talking when the rag of the maintopsail (it was close reefed) slapped hard aback against the mast from a puff of wind that seemed to make the whole ship quiver from the force it struck the masts and riggin'. At the same time it struck us. The screeching from it through the riggin' was worse than forty tomcats sending up on a still night their musical strains in concert. It only lasted for a minute and was gone again. So quiet was everything after it ceased that the order from the Captain to lay square the mainyard startled the men like a pistol shot fired close to a person unexpected. The mainyard had hardly been placed at right angles with the rails when, with a roar like a mad bull, the wind struck us square aft with such force that the man at the wheel was pinned to it for an instant, as the house over his head had an opening in the back through which the wind drew as if through a funnel. Righting himself, he firmly grasped the spokes with both hands and sung out, "Aye, aye, sir," to the Captain, who stood near him and gave the order to keep the ship dead before it.

When it first struck us, the ship's stern commenced to raise and her bows to settle down and down. Both catheads and bows were under water, and it seemed for a few minutes that she would dive into the depths of the ocean like a whale turning flukes. The bowsprit was partly buried, the jibboom and the flying jibboom just pointing clear. Hanging this way for a few minutes as if undecided to go to heaven or Halifax, she gracefully settled her stern and, raising her bows, sent half a deckload of water rushing aft to the taffrail, wetting some to the waistband of their pants who had not sprung into the riggin' out of its way. It was many long hours afterwards before our decks became clear of water again; and cold comfort surrounded us, as no fire could be lit in the galley.

The ship ran before it fine, but the wind came so hard that at times it would almost blow a man clear of the deck. The main-topsail at times seemed to be ready to leave the bolt ropes, but as it was almost new it stood.

We had run some four hours N.E. when the wind hauled more to the South. Our course now was N. to keep before it. After running a few hours on this course, the wind hauled to S.E. This brought the course N.W. By this time the Captain got uneasy and told the Mate, who with all hands had been called when the hurricane struck us, "We must heave the ship to or we shall run her ashore on some of the islands to the North and West of us."

"That's so," he replied, "we cannot do much more running now, but ride it out by the rings of the topsail sheets."

"Stand by to haul in the port main braces! Haul out the main spencer," were the orders passed to me, and by me to others, as the wind was howling too loud for them to be heard otherways.

"Hard a port," was the next order, and passed to the man at the wheel.

The ship was going through the water at the rate of about ten knots. When the wheel was put down, the mainyard, being braced up at the same time, caused her to shoot from the course she had been running. A roaring tumbling wake was left four points on the starboard quarter, showing a greenish-white track from the phosphorus in the water, mixed with the foam made by the rudder meeting with the water it was passing through at nearly right angles with the keel.

The ship had turned her head but little to the wind when

she commenced to lay over on her side from the force of its striking her side and masts, and by the time the wind was abeam she had her whole lee rail under water. It did seem that she must go over on her beam ends, and I do not know but what she would have done so if the main-topsail sheet had not parted at that time close to the eyebolt through which it was rove near the deck. Rattling through the iron quarter-block and around the sheave in the end of the main-yard it went, a stream of fire following it in its rapid course, as the sail, being freed from its restraint, was slatting itself into ribbons with reports like cannon that still could be heard above the howling gale.

The sail soon went to pieces and none too quick for the safety of the mainyard, and the topsail yard was twisted and tossed about in a dangerous manner during the sail's frantic struggles to free itself from the weather sheet and the yard that held it by the reef band. It did not take long for the sail to disappear, leaving nothing but a row of matted threads of canvas along the reef band on the yard.

A short time after that the fore-topmast staysail parted; and away that sail went before it was hardly known what had happened. A double preventer was put on the main spencer, and that sail, being new and of very heavy canvas, kept the ship partly to the wind.

The sea was blown quite smooth, but she laid well over on her side, with constant sprays of water coming over the weather rail and drenching the whole crew from head to foot. The wheel was lashed hard down, as it was no use for a man to stand there by it and be more exposed than being sheltered under the weather rail.

The men were told if they wanted to go below they could go down into the afterhatch, and be ready for a call at any moment if they should be wanted. They took the offer, and worked their way to the steerage scuttle and dove below, saying, sailorlike, they'd sooner take chances of being like rats drowned in a hole in the future, than to stay on deck and be drowned from the water coming pouring over them now. Some of the officers crawled down into the cabin. So did two of the boat-steerers. The others stayed on deck with the Captain and Mate.

Along about four P.M. the ship stood more upright. The weather was thick and hardly any distance could be seen from the ship, but as hours passed the wind dropped, so by five A.M. it was quite moderate and, strange to say, in the last hour the sea was smooth. The last feature had taken place almost at once. If such an occurrence had happened at any other time but now, it would have caused a large amount of wonder, but all hands were too tired, and had been for hours prepared for anything that might happen, to say more about it than the Captain did (he had turned in after the topsails were set), when told by the officer who had the deck that suddenly the sea was as smooth as a mill pond: "Well, I am glad of it. We can stand a good deal of that now." Then he rolled over and went to sleep again.

At four bells (6 o'clock) the Chief Mate came on deck. The wind had entirely died out and the weather was not quite so thick. A white horse might have been seen a mile off if any had been wandering around there. The two topsails and spencer, set taut by their dampness, hung without a slat. The little swell that the sea had on hardly caused the ship to rise and fall enough to be noticed.

The remains of the fore-topmast staysail and parted sheets had been cleared from the bows and stays, and were piled on the tryworks cover. The main-topsail sheets were in a heap on the main hatch. One had the ring of the clew, some of the foot rope and leech rope attached. All that was left of the main-topsail was the closereefed band fast to the yard by reef points. The ship had a dreary look; but such a change from the howling hurricane a few hours ago, to the quietness of now, was almost too wonderful for belief. No man on that ship had seen the like before.

The Mate had been on deck a short time when the 2d

Mate told him he had not started to make sail as he thought it no use until six bells (7 o'clock) as it was so thick and the men used up; it being calm, besides.

"All right, there is no need of it until the other watch is called and you have all hands to make light work of it, but you may as well send down the balance of the main-topsail and get another ready to bend in its place," replied the Mate.

MacCoy and myself, with four men, went aloft when ordered by the 2d Mate to send down the sail. We soon had it on deck, but found it a tough job casting off the close-reefed earrings, the turns and hitches had been drawn so taut by the force of the wind on the sail. By the time the watch was called we had the torn sail in a heap on deck and another topsail bent to the riggin', all ready to go aloft.

A few minutes before breakfast was called (7 o'clock), the Captain, coming on deck, spoke to the 1st and 2d Mate, asking what the condition of the sea meant, as it was something he could not fathom, to have the sea so suddenly grow smooth. The wind going down was nothing, that was often the case; but the other part beat him, he said.

"I do not know any way of accounting for it, except we are under the lee of some land," said the Mate.

"Under the lee of the Devil! There is no land for us to get under the lee of, except Eoa, and to have done so we should have been blown right over it. It blew about as hard as ever I have seen wind blow in the many years I have followed the sea, but it did not blow hard enough to blow this ship over that island and we not feel her hit the tops of the coconut trees in going over," said the Old Man. And going down the cabin stairs to breakfast (being called by the steward), he muttered to himself on the way, "Under the lee of your grandmother!"

"Whee!" said the 2d Mate. "You have set the Old Man wild."

"Yes," laughing, the Mate replied, "the Old Man is decidedly on the war path this morning." They both smoothed their faces, and with the other officers and one boat-steerer went into the cabin for breakfast. While eating, the Captain got over his spleen, as nothing more was said to arouse his ill humor.

"Have the pumps been tried?" asked the Old Man as the officers finished breakfast and started to go on deck.

"Yes," replied the 2d Mate, "when the ship got on even keel I had them started, but got them to suck after a few strokes."

When we got on deck the mist had cleared off somewhat, the sun was seen shining through it and the horizon could be seen farther from the ship, more so astern than on either beam. Someone remarked that it seemed strange it was clearing faster ahead and astern than on each side of us, where it seemed to hang in banks, looking more like land covered in thick mist than fog banks.

All hands had been set to work right away after breakfast, some bending the main-topsail, others making sail. In the midst of it all hands were startled by the Captain's voice yelling out loudly, "By G—d, here is land on both sides of us! How in hell is this?" Everybody stopped work, and, looking off larboard and starboard beam, land could be seen no great distance away. The mist was rapidly disappearing from it. In a few moments the land was perfectly clear and the sun shining brightly from a soft blue sky, with some lazy-looking small white clouds hanging here and there like pieces of cotton batting. Low down in the S.E. could be seen a hazy bank of misty matter, which soon disappeared.

The land on the port beam was about eight miles off, and it was the island of Eoa. That on the starboard beam was Tongatabu, pure and simple, and by some good fortune the ship had got in between them and into safety under the lee of Eoa, and not to windward of it and lost with all hands on the steep cliffs of volcanic rocks that mostly surround it. If we had been blown just a few miles to the North of where we were, the ship would never have been heard of. Some pieces of the wreck might have been found, but doubtful if large enough to tell what ship they came from. It was now easy to see how we had got into smooth water so suddenly.

The officers had gone up on the poop deck when the Captain sung out for the land and were now with him, looking at the chart the boy had brought at the Captain's order. They spread the chart out on top of the house, taking courses supposed to have been made by the ship during the blow. With parallel rules and dividers, they pricked off on the chart what they supposed the courses ought to have been in the ship's progress from the time the gale commenced until it ended; and they did not come within twenty-five miles of where the ship actually was. This was a mystery for the rest of the voyage.

I have often thought since of the horror that would have come over us on seeing those fearful breakers, just before we struck them. No one on board would have thought but we had plenty of sea room. No lookout would have been kept. If there had been, the flying mist and spray would not have allowed any great distance to be seen, so we would have known our fate only just as we met it. Perhaps it is best to go that way, if you have to go. But it seems to me that a horror sprung on a man like that would have been a lifetime of agony contained in a minute; and I, for one, had rather see a little longer ahead.

THE WIND came out from the West by the time all sail was set on the ship. We stood in towards the land and followed it along for a short distance, passing groves of coconut trees lining the shores back of the beaches, where openings in the dark cliffs and forbidding rocks formed small bays. The land back from the shore line was densely covered with tropical trees, vines and shrubbery. No grass land could be seen in any extent. After sailing along the land about eight or ten miles, we hove aback, some half-mile from shore, abreast a collection of nice-looking grass houses, built amongst a large grove of coconut and banana trees, back a little distance from the shore, with a fine sand beach running a short distance to the left and at right angles to a point of land that ran out from shore two or three hundred feet, having, on the side next the beach and the little bay it formed, a natural landing for boats to lie along, with sides of rock almost as smooth as if chiseled by the hand of man.

During the time we had been running for this landing, the men were employed clearing out the boats of some of their spare whaling gear, so it would be out of the way when bringing in hogs, ducks, chickens, turkeys, yams, sweet potatoes, coconuts, bananas, oranges, guavas, papayas, pineapples and shaddocks [grapefruit]. The last named fruit has every appearance of a mammoth orange. There are two kinds of this fruit. One is to the taste not quite so bitter as the other, and any person fond of bitter fruit can enjoy it. They are not quite so juicy as a good orange.

The boats were lowered, and in the boat that took the Captain was put cotton cloth (white and blue), a box of axes, powder, some muskets, lead, calico (fancy colors), fish-hooks, knives and Jew's-harps.

When pulling in towards the shore, we could see large numbers of natives, men, women and children, flocking in from every direction, some carrying on their backs bunches of bananas, others with a pole across the shoulders and a dangling bunch of chickens, strung up at each end by the legs, that made the pole bend by the weight. Again others could be seen with large hogs, all feet tied together, slung with a pole shoved between them, a man at each end staggering along; and some of the women had little baskets neatly platted from flat rushes, full of eggs or sea shells. As fast as they arrived, the loads were dropped under the shade of coconut trees just above where the boat would land, in piles each by itself.

We pulled alongside nature's wharf, making the boats fast one ahead of the other, and got out the trade goods, which the men backed into a shelter built with bamboo stocks for uprights and rafters, and covered with the broad leaves of the coconut. In this all the trade we had brought on shore was placed; and the Captain spoke with three or four of the head men, as I took them to be; for they seemed in full dress, compared to what the others were, by having the cloth (this was the most any of the crowd wore) just so it came low enough that you could not tell if it covered man or woman.

The marks of the wind could be seen on shore, in looking around after we landed. Some of the houses had been blown almost to pieces. The tops of many coconut trees had been broken off, and the long fanlike leaves of others in some places covered the ground. The shrubbery in places was stripped of its leaves and looked as if it had passed through a bad winter.

The Captain had some little trouble to get the prices current fixed, but when that was arranged, the fun would kill you. All were eager to sell stuff, those most so who had packed their trade from long distances. Unless they sold it, they would have to pack it back again and wait for the arrival of another ship.

The Captain had us open the goods. Standing in the front of the shelter, he picked from the articles brought and placed before him, holding up his fingers for as many fathoms of cloth as he would give for a hog; an ax for so many chickens; or musket, powder, fishhooks, or whatever we had to trade, for so much of this or that. They would reach for the article offered and back off. The trade was completed by some of the boys carrying it to the boats.

Now and then two or three would dump their stuff down in front and it would get mixed together, or the Old Man would get himself mixed and think he was buying one man's when it belonged to two or three, then pay one for it and have the others object to having the man take it until they were paid for what belonged to them. It would take the head men in full dress some time to explain the matter by rushing amongst them with a stick and cutting right and left. Order being restored, trade would go on again. Sometimes, when measuring the cloth, one fellow would slyly catch the end hanging down and give it a pull, to have it slip through my hands and get a bit more.

I was chosen by the Captain to measure, I suppose, on account of my arms being the shortest. He almost always had me do it, and not any of the other boat-steerers; and when a native tried the pull on me he never got a bit more than belonged to him—I looked out for that. Sometimes I would call the attention of our leading full-dressed party to the act, and the fellow would get his extras with a rap that would make him howl. By dark we had the boats loaded as full as they would hold.

We stood off and on with the ship during the night. After breakfast the Captain took our boat, and with some articles of trade in it, started again on shore. The 4th Mate (who had taken the place in our boat of the 2d Mate, who was in charge of the gang to fill the water casks) and myself, after the Captain had stopped trading for a time, took a stroll amongst the houses a short distance back of the high-water mark. The natives seemed very pleasant and kind, the women laughing and jabbering like a lot of monkeys, and using about as much gesticulation, and affording us a good sight of their agility and well-formed limbs.

The only dress they had on was a piece of cloth around the hips, fastened by shoving one end in the fold next the body. It did not take them long to prepare for a bath or bathing. They might have had on more, but certainly could not have had less, to fulfill the scriptural text. It answered the purpose, however, but it would not have done the business if it should shrink half in fore and aft. Still, they were all right; and when one (with her sparkling black eyes, well-shaped light-yellow body and limbs, and quite pretty face, if it was dark) came up to the 4th Mate with some ripe bananas and wanted him to take them, he placed his arm around her; but like a cat she slipped away from him.

Running off a little disance, she turned facing us with her roguish black eyes sparkling with fun, and laughingly called him "Devil! Devil! Too much Devil!" in plain American talk. She, like all of them, had been taught by the missionary element to call and treat all white men as such, if they did not belong to the church. This had been so impressed on their simple minds that I have never heard of but one instance to the contrary, in that whole group of islands. Of this I will speak simply to show how really the true principle of religion has no depth with them. This instance I speak of happened to a sailor, but has nothing to do with the missionaries, black or white, amongst the group (for a tale could be told that would astonish the dear demure sisters and brothers of the Home Board, if only a little bit of it is true).

The man I speak of ran away from a ship that touched at one of the islands. He was a convict in Sydney, smuggled himself aboard a whale ship laying in the harbor on the eve of sailing, stowed himself away in the ship's hold, and never showed on deck until the ship had been two or three days at sea, too far for the Captain to turn back and land him. This ship he left at the first chance, and, as he had come on board in such a way and was a bad man anyway, the Captain was glad he left. We spoke the ship that the white missionary, on the island, later got him taken on board of, and they told us the missionary had said that the man had almost played havoc with his Christian people; and he begged the Captain to take him away. The fellow did not want to leave, but force was used to put him on board.

It seems, after the ship he ran away from left the island, he came out from his hiding place and went to the white missionary's house and told him a pitiful tale of how much abuse he had suffered; that he had been a bad man in years gone by, but now wanted to live a better life. No doubt he told this purser's story to destroy any bad feeling against him from anything that the captain might have said.

The parson gave him some clothes and a house in his yard to live in, no doubt thinking he was gathering in a lost lamb. As the fellow picked up the language he prayed the harder, so, by the time he had been on the island six months, he was a leading member of the flock. Not longer than a year had passed before two or three half-white children were born. Such a strange affair caused the head of the spotted lambs to make a few inquiries; and he found he had better amend his by-laws by adding a clause in the Devil section to have it include all foreigners, whether belonging to the church or not. The fellow had persuaded them that, as he was one of their number by belonging to the church, it was all right. The fellow said he never had such a good time in his life. The hardest work about it was when he had to do the praying acts. He also told them, on board ship, that he had broken the rule in no other way as a member of the church, except in getting found out.

Rambling around for a short time, and being called Devils a good many times, we steered back to where the Captain was almost unshipping his jaw, trying to talk Tonga language enough to give an old deacon-looking native Hail Columbia for telling him he was "No good do all same." A young fine-looking woman was standing a short distance off, laughing and making faces behind the old fellow's back, which a number of young girls seemed to enjoy.

"Ho, ho!" says the 4th Mate, catching sight of what was going on. "I guess the Old Man has placed himself in a position to be called a Devil, by the look of things."

The Old Man did not see us until we were close to him. Then he turned to us, red in the face from anger, and pointed at the Deacon, saying:

"That psalm-singing, knock-kneed, crooked-legged, wall-eyed, antiquated representative of an Egyptian mummy, who has been dead for the last twenty years but don't know it, had the cussed impertinence to take me to task for offering that girl three fathoms of calico for a kiss. And she would have done it if that picture of a Chinese idol had not caught her by the arm and pulled her away, the son of a sea cook!"

The old native, by the time the Captain had finished, was out of sight. The girls, women and young men seemed to enjoy the fun, but did not dare to show how much openly for fear it might be reported to the elders and they would suffer for it.

After everything was put in the boats, we went on board. The boats were hoisted up and all sail set. By dark we had stood out clear of the land, and we set the course N.E. for the Line. All hands went to supper except three boat-steerers and the man at the wheel.

The ship had a strong smell of fruit around her, but a stronger one came from forty or fifty hogs running around decks, and a hundred or two turkeys, ducks and chickens in coops, tied by the leg or loose. Some fun for the boys was made by the hogs sliding across decks, as the ship lay over in the first strong wind we had after taking them aboard. The hogs for a time kept huddled close together, and like sheep, "as one went, all would follow." Sometimes one would take fright, by the first man of the watch starting aft quickly to obey some order that called for the whole watch to do, and the other hogs, starting to follow perhaps square across deck, would get mixed amongst the men's legs and tumble them over. The men knocked down would strike out with fists and feet, joining the chorus of oaths showered with kicks and thumps on the poor swine by the others who kept their feet and were trying to struggle through the squealing herd.

After a few days they got used to each other and only occasionally got mixed. The hogs, though, never would learn to rest quietly in the lee scuppers where they had been slid from the weather side, but would dig their toes into the deck and struggle to windward again, the ship not lying over quite so far at times as others. They would crowd as close to the weather bulwarks as possible, some lying down with their feet braced to hold them, and give a grunt as much to say, "There now, I am going to stay right here, no more sliding for me." But it would be no use, the old ship would give a little twist and miss a half-step to windward, and, as though mad for the act, raise her weather side high, bringing her decks quick to the angle of forty-five degrees, and send the grunters tumbling, rolling, sliding with squeals loud and shrill enough to make the jewell blocks on the topsail yards rattle. Woe betide the fellow that was crossing fore and aft in the line of that avalanche of hog meat. Sometimes a man, so caught and rolled with them into the scuppers, would come out of it looking as though he had been run through a rag-picking machine—a dirty one at that.

The taste of fresh pork, to those who have been living on old horse for some time, is fine, but the trouble and nuisance of having hogs run loose around the decks is almost too much for the pleasure of eating them. The decks have to be washed, at such times, two and three times a day. All through the day one or more of the "jimmy ducks" (poorest men amongst the crew) are running here and there with brooms and shovels. Even then you have to be careful when heaving down on deck coils of braces or running riggin'.

There was but little salt meat cooked for some time after

we left Eoa. Pork, turkey, ducks, and chickens were the bill of fare, with sweet potatoes, yams, and taro on the side. We all waxed fat and hearty.

THE WIND was fair and the weather pleasant during our run to the Line. We used up a couple of weeks or so making the passage, and struck the Line in about the Long. 165° West of Greenwich. The season for whaling on the Line is all the year round, so it does not matter much what time ships cruise there. Some ships cruise there the whole voyage, only leaving to go to the North or South for recruiting ship and giving the men a run on shore once a year. With ships that cruise on New Zealand and the Line also, May, June, July, August, and September are the months chosen for the Line cruise, as these are the most stormy months on the Southern Grounds.

Ships coming from the South try to strike the Equator as far to the East as they can on account of a strong current most always setting to the West; so much so, that ships cruising there keep sharp on the wind, which prevails from the East, and, doing the best they can to work to the eastward are drifted to the West, sometimes twenty to forty miles in each twenty-four hours. Those who cruise the voyage there, when carried by the current as far West as they care to go, stand out to the Lat. of from 5° to 7° North or South, generally North, there get into variable winds and work the ship to the East as far as wanted, then strike the Line, and go through the same process as long as the voyage continues.

On our passage to the Line we sighted a number of islands, but no whales. We had been cruising for a week or so without a sight of a whale, and it was getting dull work. One day the officer at the masthead hailed the deck and reported, "A sunfish in sight, short distance from the ship." The Captain told the Mate that if he wanted to do so, he might lower his boat for him. The mainyard was hove back and his boat lowered. The fish could be seen a short distance away from the ship by his dorsal fin sticking now and then a foot or two above water. As they seem to have no sense of fear, he pulled up to it and the boat-steerer darted an iron into it. The thing made but little resistance and he towed him alongside. A hole was cut through the thin part of his body, through which a stout strap was put and hooked on a six-fold tackle which had been sent aloft for that purpose. The fall was taken to the capstan and he was hoisted on deck.

These fish [*Mola mola*] are queer customers, nearly oval in shape, the body looks as if it had passed through a set of rollers that had flattened it by tightening down more closely after half the body passed through from the head, and got tighter as it went on, until the hind part was pinched off, leaving a thin ragged edge almost straight up and down. The forward part of its body is about six to eight inches through, and is thickest in the center just back of the mouth, which is a pinched protuberance, midway in the pancake shape of the forward part of its body, and is quite small and almost round. The jaws are cutting edges of a hard bonelike substance, instead of teeth. The eyes are small, one on each side, a short distance back of its mouth. The body is about five or six feet across, and has not a bone in it; of gray-black color covered with a coarse rasplike skin, similar to a shark's but much more harsh, under which is a firm white meat or substance, whichever you might call it, having something the appearance of gristle; but it has no grain. When cut with a knife, it leaves a smooth shining surface, as would be in slicing a piece from a candle. By cutting a piece into the shape of a ball and bounding it on deck, it has almost as much elasticity as rubber.

A part of the inside, near where the nape of a Christian fish would be placed, is another kind substance, composed of broad stringy matter. This can be eaten, is tough and has a rank, dry taste. Between it and the taste of shark meat, one can hardly tell which is the worse. The dorsal fin is long and pointed, standing stiff upright, commencing about 18 inches forward of the after part of the body and standing about two feet in height. The lower fin is the exact counterpart of the upper, and sticks directly downwards. Their food seems to be the little oblong flimsy creatures that at times are seen by the thousands on the surface of the water in light winds, and the bladderlike thing called by sailors the Portuguese man-of-war.

When landed on deck, the sunfish looked as if he would weigh at least fifteen hundred or two thousand pounds. His length was about six feet and width almost the same. We cut him open and took out his liver, which would fill a bushel basket, to save the oil in it, as it is very fat and makes almost as much oil as the bulk of it.

The only way to save the oil from the liver is to let it stand in an open tub exposed to the sun until it rots and becomes putrid. The stinking mess, after a time standing in the sun, will clarify itself, and the oil can be poured off, leaving a very little red muddy-looking matter at the bottom. This oil is the finest in the world for softening and preserving leather; and those that have tried it claim that it will cure rheumatism. It has the bright rich-looking color of pale brandy. The carcass was tumbled overboard and the decks washed off, but the smell of the fish on deck lasted for a week.

A FEW days after our catching the sunfish, one fine forenoon, the ship, with all her sails set, was sharp hauled on the gentle wind that made the canvas stand swelling out from the yards and sheets like the bosom of a fair, fat and forty dame in full dress. On the sea, here and there, now and then, was to be seen a small lazy break, which showed milk-white in contrast to its blue surroundings that rivaled the azure above. The ship's sharp bows were parting the water beneath them with just force enough to send clear of the shining copper that marked her water line a thin white roll of foam that soon turned into bubbles and went rippling along aft with a tinkling sound.

The watch on deck were at various jobs, some at work in the riggin' putting new chafing gear on it in place of that worn out by the yards or ropes coming in contact with it, some seizing on rattlins that needed replacing, some tarring places in the standing riggin' that had been bleached by the wind or had had the tar chafed off. Some were braiding sennet, some making spunyarn; the officer of the watch and Mate here and there to see the different work properly done. The Captain, leisurely walking the weather side of the quarter-deck, humming some tune, was now and then casting his eyes aloft to the sails. Catching sight of a wrinkle in the weather leech of the main-topgallant like a boy's face beginning to cry, he would sharply speak to the man at the wheel: "Too near!" This would be about the only sound of voices heard except now and then an order from the officers to some men at work aloft.

But soon every man stopped in the work he was at, and all listened with sparkling eyes and pleased faces. The Captain stopped abruptly in his walk, and with an eager look turned his face upwards to the main-royal yard, as from the officer and boat-steerer leaning over it, on the lookout, came the pleasant and welcome sound of:

"T-h-e-r-e s-h-e b-l-o-w-s! T-h-e-r-e s-h-e b-l-o-w-s!"

"Where away?" says the Captain.

"Two points on the lee bow, sir. T-h-e-r-e she blows!"

"They are sperm whales, are they not?"

"Yes, sir, a small school."

"How far off?" again the Captain asked.

"Two miles and a half. There she blows!"

Then came the orders, sharp and fast: "All hands come out the riggin'—put away marlinspikes, serving boards, spunyarn and mallets, tar and slush buckets! Call all hands! Get the lines in the boats! Haul up the mainsail! Haul aback the mainyard!"

Well, if it was not fun to see those decks now that had been so quiet a few minutes before; men with tar buckets slung to them sliding down backstays that they had been at work on, others throwing down rattlin stuff and spunyarn out of the riggin' and jumping down on deck, hastily grabbing up their different material and rushing with it to the places where it should be put—some in their haste running into each other—but all wild with excitement.

The mainyard was hove aback and the boats hoisted and swung, with strain enough taken on their tackles to raise them from the cranes so these could be swung on hinges parallel with the rail, all clear to lower.

The whales had gone down and we were waiting for them to raise, to see how they were going before we lowered. In about twenty minutes they came up about the same distance off, and four points off the lee bow, showing they were still and working to the leeward. "Fine chance at them," the Captain said, as he gave the order for the four boats to lower.

After getting clear of the ship we set our sails, the men took paddles, and away we went, the boats skimming over the water with hardly a pitch or roll, as the sea was like a smooth lake. Before the whales went down again we had gained so much on them that they could be seen plainly from the boats.

Shortly after the whales disappeared, the paddles were laid into the boats, but we all kept sailing towards the spot they had been seen in when they went down. We had run about twenty minutes when we saw the Mate's men paddling like a lot of Indians, and at the same time we saw the whales right ahead of us, not more than a short quarter of a mile off. But, as we were on the eye of the whales, we had to lie still. The Mate's boat was so far to the windward that they could come in behind them without being seen.

Soon the Mate's boat-steerer stood up, and shortly after he darted his irons. White water flew in every direction close ahead of him, and a minute after his boat, with the sail flapping violently, could be seen tearing through the water, the bows buried deep in two sheets of foam, one flying from each side.

"He's fast! He's fast!" was the glad shout from each boat, as we rolled up our sails, took the oars, and stood by for a chance to strike one of the other whales when they came up; for all had disappeared when the Mate struck his whale.

In a few minutes up came the whale that had been struck, and with him three loose ones. They huddled close together, striking the water with their flukes and shooting their spouts out with an affrighted noise that made things hum.

"Pull ahead!" came the cry from each boat-header, and away dashed the three loose boats for the center of the commotion, coming into the raging waters about the same time. All three boat-steerers were called to their feet, and with a firm grip on the irons, as each one came near enough to strike a whale, he raised and darted.

There was something going on in the next few minutes that would discount for yelling any madhouse in the universe. This state of affairs was brought about by the whales taking opposite courses to each other, which caused the lines to cross. Not getting them clear quickly enough would send the boats below in short order, or at least lose lines and whales.

It would be hard to tell all that was said at the time, but some of the boys swore afterwards that two or three of the officers could not speak loud enough to hail a man from deck who was standing abreast the fair leaders in the main riggin'. Some oars were broken, hats lost, fingers bruised, arms skinned, a board cracked in some of the boats, and more or less line run out before we got out of the biggest snarl anyone in the boats had ever seen before in a school of whales.

The whales showed good play after separating, and we soon had them turned fin out. They all went into their flurry and turned fin out within a mile of each other, and the ship, being to windward, ran down, taking the one to windward first and the lee one last. Soon we had them fast alongside, the tackles were got up, and, after eating dinner and supper together and getting the light sails, jibs and mainsail furled, all hands went below except what's called a boat's-crew watch, which is composed of six or eight men. The ship lay with her fore-topsail yard on the cap, her reef tackles hauled two blocks, foresail hauled up snugly, and the yards laid square and aback; and as the main and mizzin topsails were braced full on the starboard tack, spanker set, and the wheel lashed hard aport, the ship would keep head to the wind and make her wake dead off the weather beam.

At the first crack of day the next morning all hands sprung from their berths at the cry of: "All h-a-n-d-s a-h-o-y! Tumble up and man the windlass!"

The stages for standing on to cut the whale had been put into place the night before. The officers who cut were on them and had made the proper cuts to receive the blubber hook and called out, "Overboard hook!"

The hook was dropped over the side and shackled to the immense lower block, which is a counterpart of its fellow lashed to the head of the mainmast above, each having two sheaves and a fall rove through of three-and-one-half Manila rope of the finest quality for strength, one end fast to the block above and the other leading forward down to the windlass. (As there are two sets of these blocks, one fall leads to port and one to the starboard side of the windlass.)

It was my "overboard," as it is called. Each boat-steerer has to take his regular turn as whales are caught, and it makes no difference how small or large the whale, or the number taken in any one time; it is his turn until those are cut in. Sometimes one fellow will have a fifteen-barrel whale that he would not go overboard on more than once or twice, and the next fellow will have a hundred-barrel one that he might be over eight or ten times. He need not look for help in other directions, for none can he expect from his fellow-mates.

I had been ready with the band around my body, attached to a long light rope. One end of the rope is passed through a grommet, or hole, in each end of a folded piece of canvas that would almost reach around a man's body, four or five inches wide. It is then brought back to the standing part and fastened with a splice, leaving a loop to allow the rope play enough through the grommets for the band to open wide enough to slip over the head and shoulders and under the arms. Then, by hauling the rope taut, the band is drawn tight around the body and kept in place under the arms, leaving both free to act. The name of this is the "monkey rope."

I stepped to the gangway, out of which I sprung. Landing on the slippery side of the whale, I was only able to keep on him by help of the man tending the monkey rope and the little help got from catching my feet into the cuts made by the spades around the piece to be raised. Next, I had to place the hook in a hole cut for that purpose near the end of the blanket piece.

The hook has a rope fastened on the back of it, close down to where it starts to turn, that is long enough to reach a man on the whale. I got hold of the rope, braced myself as firmly as the cuts in the whale would permit, and the fall of the tackle was overhauled, letting down the lower block and hook. As it came down I had to pull the whole business towards me on top of the whale. If it should slip off from the whale before I could haul it towards me far enough to catch hold of the hook and place it in the hole cut for it, down it would go rattling, hook, block and all, between the side of the ship and the whale; sometimes taking fifteen or twenty minutes to clear, when the whale happens to be rubbing the ship's side close. At the best, at such times, a heap of tugging is needed to round up on the fall and get the hook in place again for another attempt. The black looks and words the poor devil on the whale gets at such times might cause even a mule driver to shudder.

It is no easy task to haul a hook that will weigh about one hundred and twenty pounds, the weight of block and fall besides, some fifteen or twenty feet out from the side of the ship and turn the hook, point out, in the hole, taking into consideration that you are all the time clinging like a crab to keep yourself on the whale's side. Sometimes when it is rough and the whale heaving with the sea up and down, the boat-steerer will be swept by the force of the sea from the whale into the sea, outside, ten or twenty feet, right among the sharks that are always around when cutting sperm whales. The man who is tending the monkey rope at such times has to pull quick on it to get the man back before the sharks sample his legs or other parts of him. Then again, if the sea should wash him inside the whale and the rope is not pulled quick, the man may be crushed between the ship and whale.

By good luck I soon had the hook into place, holding it there by standing on it the best I could until a strain was taken on the fall, so it would not slip out. Then came the merry rattle of the iron pawls, and as the windlass rolled around the piece soon was lifted clear of the carcass. As this was all I needed now overboard, I came on deck. Taking off the monkey rope, I joined the waist gang to carry on my share of the work there.

The boys were full of spirits, and the way they made the old windlass roll caused the cry of "Board oh!" to come often as one blanket piece after another swung over the gangway. The boys at the windlass were singing the soul-stirring, heart-rending songs: "High, Randy, Dandy," "Oh Off She Goes, Off She Must Go," "Jigger in the Bum Boat" (or words to that effect), "Sally in Our Alley," and many others.

The heads were hoisted on the deck whole. As not one of the whales would make more than thirty barrels, by dinnertime we had three of them cut in. After eating, we started cutting the last one, and by some misscut, when half-finished, the piece tore off. The wind had freshened up and there was quite a sea on. The whale alongside made quite a fuss with the swell breaking over him, which now and then would send a shark on top of the carcass, as it settled below the surface, and left the blue staring-eyed monster high and dry as it raised out of the water again.

By the piece tearing off, the whale was clear of the ship, except his flukes which were fast by the chain around them that led through the port in the bow. Slipping the band of the monkey rope over my head (the rope being taken by one of the others), I made a spring from the forward stage onto the carcass. The tackle was overhauled down the side, with a hook on it. A long, slightly curved piece of hickory, pointed at one end and flat at the other, with a hole large enough to take a small-sized rope, was passed down to me. I was to shove this crooked wooden "needle" down through a hole that had been cut through the piece of blubber that had torn off and was floating on the water, some ten or fifteen feet of it, with the other part fast to the whale's carcass.

After one or two attempts I passed the needle through the hole and handed it back to a man who had got down the ship's side to reach it from me. This being passed to the deck, three or four men took hold of the rope the needle had been threaded with, and commenced to haul in on it. In the meantime a rope of a large size, fast in a chain strap, was bent on the needle rope and this was hauled through with half the chain strap. This being double, I had to hook the two parts over the hook; and by so doing the piece could be raised again and hove high enough to get a proper hold of it.

Everything had worked very well until we got that far. Then we were bothered by the whale's swinging away from the ship farther and farther until he was almost at right angles with her, leaving me perched on the carcass like Robinson Crusoe, "monarch of all I surveyed." But different again from him, I had plenty of company, which he had not; and no doubt he would have wished, as I did, for less of it. The water was swarming with sharks. Now and then one would rush across the carcass as it settled below the sea, taking a bite out of it, and leaving a saucer-shaped hole six or eight inches across, cut as clean as a sharp knife would have done. At times they would come very near to me, but fortunately none happened to take me in their course.

The ship's headyards were filled away and the jib hoisted. The wheel was put hard a-starboard, and, on her falling off from the wind and gathering way, the whale and orphan came alongside. I had kept the strap in place without much trouble except at such times as the seas would keep my head covered longer than usual. I hooked on the chain almost the minute the whale came alongside, and the piece was soon crawling aloft.

When the whale first began to go off from the ship, the boat-steerer tending the monkey rope did not think the whale would swing away far, so he neglected to bend on another piece of rope to the one he had attached to me. The whale, still going farther off, brought the end of the line in his hands, to which he hung on and was dragging me off of the carcass into the water, which was as full of sharks as some place is full of sinners. As I was hanging with all my hands were capable of doing to the chain strap through the hole, I yelled out to him to let go his end, which he did—though not too quick. In a minute more I should have been flopping in the water, with more sharks around me than I would have had any use for. Perhaps two or three men, pulling quickly on the monkey rope, might have been able to save some of the upper works that belonged to me; but as I had use for my kickers, I objected to the trial.

The Mate sung out to lower a boat and pick me off the whale, but I told him I could hold on all right and make a landing if I did not run out of provisions. "Look out you do not slip off into the water," was his reply.

I felt perfectly sure of my position after getting both my legs down through the meat that had been cut with the spade in scarfing the blanket piece that was taken off before. This had been cut deep enough to strike the ribs, between which I worked my feet, and it would have taken quite a pull to dislodge me. I came out all right, but I do not care to have another trip like it.

The amount of sharks that will gather around a dead whale shortly after it is killed, in the tropics, is wonderful. For days, cruising around where you know sharks abound, not one may be seen, but strike a whale and get him spouting blood, and then you will find them coming, from where you know not.

By dark we had finished cutting, sent down the cutting tackles and washed decks, and all the blubber was stowed in the blubber room. The topsails were mastheaded; foresail, jib and spanker set. During the night the heads were cut up and the junks cut up and put in junk casks, which were lashed along amidships. The case oil had been put into the try pots, ready to start the works.

At daylight the fires were lit under the pots and soon a column of black smoke was pouring out the back arches almost halfway to the mainyard and drifting away on the lee quarter in graceful curves. On the third afternoon from the day we started to boil, all there was left in sight of the four leviathans was in the ten or twelve rows of casks lashed to the rails each side the quarter-deck. As these were all cow whales, they only made about one hundred and fifteen barrels.

As soon as the oil got cold (it takes two or three days in the tropics, and oil should never be stowed in the hold when warm, for it is very apt to make the casks leak), we stowed it below in the lower hold.

The next day after stowing down the oil, the whole day was devoted to scrubbing the rails, paint-work and decks; and when all was done you would have found it hard work to soil a white pocket handkerchief by rubbing it on any part.

Cruising Down the Line

A FEW DAYS after we stowed down the oil, a sail was reported from the masthead, two points on the lee bow. As we were standing on opposite tacks, we soon came abreast of each other. The stranger, when four points on our lee, hauled up his mainsail. This, amongst whale men, is an invitation to have a gam (or to speak the other ship).

When our Captain saw this, he gave the order to haul up our mainsail, to show we accepted the offer. Standing along until the other ship bore about two points forward our lee beam, our wheel was put hard up and the yards squared. As the ship got before the wind and pointed her flying jibboom for his mainmast, he at the same time hauled aback his main-topsail, and being not far apart, we soon came within speaking distance of each other. Running across his stern, we found this to be our old acquaintance the *Mitchell*. When asked the question, "What success?" he said in reply, "Six hundred and fifty barrels of oil since seeing you last." This caused a feeling of sadness among the whole crew on our ship, not because we envied him his good luck, but to think we had been so unlucky.

On the invitation of Captain Sullivan for our Captain to come on board, we rounded the ship to, under his lee, with our mainyard aback, on the same tack he was laying, and lowered a boat, and our Captain went on board. The boat shortly returned to our ship with the chief officer and a boat's crew from the *Mitchell,* for a day's gam. After hoisting our boat up we braced forward the main-topsail and kept the ship off until we had separated two or three miles, to spread our chances of seeing whales. When this was done, we hauled up and stood along together.

As a rule, when gamming, where the captain of one ship visits another, the mate of the ship visited, with his own boat's crew, returns on board the ship the captain leaves. Sometimes where there are two or three ships gamming and the captains go to one ship, the mates to another, the second mates to another, lots of fun is had during the time. In fine weather, at such times, half the night is passed in singing, dancing on the decks, and spinning yarns.

The two mates talked over what had taken place since last we met. They had seen whales quite often and those taken were all large size, one of which made them one hundred and twenty barrels of oil, so the difference was not so great in the number of whales taken. They had experienced the same hurricane we were in, had bulwarks and boats stoven, and lost the flying jibboom and some sails.

About ten P.M., a light being set for their mate to return, he went on board. Our Captain shortly after came back, telling our Mate that we would keep company with that ship for a time. The mainyard was braced forward, and all hands but the watch went below.

In company with the *Mitchell* we cruised for about two weeks, the ships working to the westward from the force of the current. During the time we had seen no whales, but every few days the captains would pass evenings together. We were now about on the meridian of Greenwich and

Lat. 2° S., no great distance from the S.E. island of the Kingsmill Group (or, as called by some, the Gilbert Islands). In conversation between the two captains, when the subject of allowing the natives to come on board the ships was discussed, Captain Sullivan said he had no reason to stop them when his ship was near some of these islands; but around some of them he had no doubt it would be dangerous to allow any on board. Our Captain, who had for many years cruised among the different islands of the Pacific, including the Kingsmills, and knew pretty well what islands to trust, called the names of several in this group that might be trusted, among which he mentioned the name of the S.E. island, Byron Island, the one we should sight first.

One forenoon the ship with all sail set was heading to the S.W., going through the water about three or four knots. The blue color of the sea was so intense that a washtub prepared for bluing clothes would have been nowhere. The sun was shining brightly out of a sky so clear that it would seem a thousand miles could be seen into it by looking above in any direction.

The 2d Mate and myself had the lookout at the masthead, both leaning over the main-royal yard (the sail was furled), one each side of the royal mast, with feet placed on the topgallant crosstrees. The *Mitchell* lay hull down to the windward, head along by the wind on the same tack and showing that she, like ourselves, could not catch sight of the low bushy spouts. The 2d Mate and I had been spinning yarns about some scenes of home, when I saw him gaze steadily off the lee bow for a few seconds, then reach his hand back of the mast and take out the spyglass from a box fastened there to hold it, at the same time saying, "I was just going to sing out, 'Sail Oh.' " He put the glass up to his eye and looked steady for a minute, then dropping it over his arm on the yard, asked me, "Do you not see something looking like a sail two points on the lee bow?" After looking a few minutes I could see what looked like three or four ships. "Well, I do," said I, "but there are six or eight, seems to me, very near together. What does it mean?" He burst into laughter, saying, "Here, take the glass, and look at them. See what you make out of it."

Putting the glass to my eye for a minute, I could not help saying, "That beats anything yet, trees growing out of the ocean!" That's what it looked to be through the glass. This was the first time he or I had seen what became familiar to us afterwards. The trees we saw were coconut, growing as they do in that low latitude, to perfection, reaching often with their tuftlike tops one hundred and twenty-five feet from the ground without a limb, leaf, or branch until the bunch of graceful, long, waving, fanlike leaves or branches spread out from the pipe stem of a trunk, which carries almost the same girth from the ground to its top, where the coconuts grow in clusters.

What we could plainly see through the glass was the trees two-thirds of their length, but the land they were growing out of was so low it could not be seen until on nearer approach. This island, like many other atolls, was but eight or ten feet above the level of the sea. Most of the kind are circular and inclose a lagoon, with a passage into it, through the coral reef and sand, that will in some of them allow quite a large-sized vessel to anchor. As a rule, outside the reef that compasses them the water deepens abruptly, and no great distance from the line of breakers that surrounds them, 100 fathoms can be run out and no bottom be found.

There are many theories accounting for their formation. The one I think the most reasonable is that the coral insect has built up the foundations from the rims of sunken volcanic craters. Of course, that the insect cannot work above water is a well-known fact: but when the surface of the ocean is reached, pieces by the force of the sea are broken up, some ground into sand, and, drifting matter being caught on the inside edge, in time is formed a barrier that only the highest wave would wash over. The coral insect, being unable to build seaward on account of the depth of the water, extends his work on the inside; so in time a place is formed that some drifting coconut can lodge on and take root—and so the job would be done.

The land, or trees, being reported by the 2d Mate to the deck, the Captain said in answer, "All right, that's Byron Island. You will shortly see the yellow-bellied beggars coming in their canoes by the dozen."

We had not stood towards the land more than an hour longer before we could see thirty or forty canoes. Some had sails set, standing towards us. Others had not but were paddling partly across our bows to intercept us. They had five to six persons in each canoe; more women than men, we found, on their approaching nearer.

By the time dinner was called, the land was plainly in sight, a low bank laying on the water like an immense raft, with hundreds of sticks standing upright at different angles, the numbers and height of which gave it a topheavy look that was added to by the wide-spreading fanlike leaves on top. The natives climb to the tops of these trees through the day and watch for the appearance of a ship, which from that high elevation can be seen almost as soon as a ship can see the island. Upon a ship being sighted, the canoes are hurried to the water.

Some coconuts, shells, fish, and sometimes the sap of the coconuts are brought off to exchange for tobacco. This above all other things they crave. Money they spurn, bottles they will take next to tobacco; but they care for little else. Some of the islanders in the group braid the finest kind of hats and mats, some of which will take them days to make; yet these can be bought for a piece of tobacco that does not cost in the States over one or two cents.

The sap from the coconut trees, when first taken down in

the morning before the sun warms it, is one of the most delicious drinks I ever tasted, but it will soon ferment after becoming warm, and in a day or two will produce intoxication and afterwards a roaring headache, so I have been told. It was enough for me to hear about it, without trying it. As both kinds of the sap are brought off to the ships in like containers, empty coconut shells, if a lookout is not kept by the officers to see what kind comes on board, some of the men will soon get in a way that it would be impossible for them to say "Truly rural," twice running, without unshipping their lower jaw.

After dinner the mainyard was hove aback and the foresail hauled up. The island was about two miles off, right ahead, and by the way we were set to the leeward by the current, the ship with her main-topsail to the mast would forge about enough ahead in an hour or two so that we would come under its lee about half a mile clear of the gleaming line of snow-white surf that was breaking on the edge of reef that surrounded it. The other ship had run off with her yards checked in and was hove aback a mile or so to the windward of us.

Before we went to dinner, some of the canoes had made out to catch on the ship's side and, by having some ropes thrown them, the natives scrambled on deck. One watch with an officer had remained on deck, while the rest got dinner. When the mainyard was thrown aback and the ship's way deadened, the other canoes, that had been lying off watching how the first lot were received, came shouting and paddling alongside, with great glee and pleasure. Soon we were three deep with canoes on the lee side, and others who could not get in there had piled in alongside to the windward.

In a short time our decks were full. A rope was stretched across from the mainmast to the rail on each side, and officers were placed on guard to allow no native by, except now and then when one would have something to trade that the after folks wished. By this means the quarter-decks were clear, but from amidships to the bows was one mass of shouting, laughing, happy natives around the men, offering for a small piece of tobacco anything the sailors would take. The other ship seemed to have as many canoes around her as we had.

These natives are quite pleasant-featured. They have a yellowish-colored skin, but constant exposure to the burning rays of the sun makes them look darker than they would otherwise. Some of the women, for such kind, are quite pretty. Now and then one can be seen with a faint tint of red showing through her dark cheeks. The men are lightly clad by having a narrow strip of some kind of fabric wound once about the body and one end brought up between the legs and caught in the standing part around the front. A good many of them did not have even as much dress on as that. The women had a fringe of rushes, split into shreds and fastened to a string, tied around the waist. These hung over the hips down halfway to the knees, and, with the care they used, afforded proper protection. The young girls had pretty forms, as a rule.

The canoes showed masterpieces of work in more ways than one, as they were built of small, short lengths of different kinds of woods. They were finely shaped, about fifteen or twenty feet long on top, and sat, when loaded, with the gunwales about twelve or fourteen inches above water, as graceful and buoyant as a duck. On one side two long pieces of light timber projected out some eight or ten feet, the ends of each lying across both gunwales about ten feet apart. The ends on the gunwales were firmly lashed to both. On the outer ends were sticks of timber of some light, floating wood, fashioned like sled runners. These lay on the water parallel to the canoe, fastened by lashing the projecting ends two or three times. This was to prevent the canoe from rolling over.

The hulls of the canoes were built of small, narrow strips

of wood, neatly fitted together and held in place by being strongly lashed, piece by piece, to each other with the fiber of coconut husk twisted into strings and run through holes. When done, they showed as pretty a model and finish as though made from one piece of wood. The seagoing qualities of these canoes are fine; and with their sails, made of mats, spread in a good breeze, they will fairly fly through the water.

By four P.M. the ships had drifted four or five miles to the leeward of the island. All of the stuff they had brought on board had been disposed of, but still they seemed to be in no hurry to leave the ship. The Captain told the Mate to brace forward the mainyard. This being done and the foresail set, she was kept off four points from the wind. This caused the old ship to raise a white bone under her bows, and set the canoes, huddled together alongside, to smashing and crashing into each other and against the ship's side in such a manner that in a short time there would have been no whole ones left. Seeing this, the natives on deck began yelling and rushing fore and aft, some jumping overboard from all parts of the ship, those in the canoes casting off the lines that held them and adding their frantic yells to the others'. The ship's wake for a mile astern was seen covered with canoes, some bottom up, others half-sunken, among which were men's and women's heads bobbing up and down with one hand raised high, holding their precious tobacco to keep it dry, as they swam like fish for their canoes. It was rather a rough way to use the poor devils, but it did not take long to clear the ship of them.

After we had the decks clear, we wore ship to the N. & E. on the same tack, the other ship having gone a short time before. Men were put over the side to draw water and the scrub brooms set to work washing decks. The mess to clean was almost as bad to handle as a barnyard. In about an hour the job was done and we went to supper. All hands seemed to have enjoyed the fun, some showing the hats and mats, others shells, green coconuts, fish lines and other things they had bought.

I spoke to the 2d Mate, during our watch on deck that night, about the canoes that were upset, and said I was afraid that they might lose them.

"Fiddle-de-dee," he replied, "lose nothing! In the water those natives are like fish. They can bail out one of those canoes, when swamped, in a very few minutes, by getting to one end and bearing down on it. By so doing they raise the other end as high out of the water as they can, give it a shove from them, and let it go at the same time. This will send half the water out of it; and doing this once or twice will so relieve it, they can get into it and bail the balance out, pick up their loose stuff floating around, and think no more about it than a person would about taking a bath."

For a couple of weeks we cruised in company of the *Mitchell,* but saw whales only once, and took one. He made sixty barrels, one half belonging to the other ship, as we were mated. That is to say, the ships agree, as long as they are in sight of each other, that both ships' boats are to help each other as belonging to one ship; and the oil taken during the time is to be equally divided between them.

One morning after breakfast the *Mitchell,* on our weather bow about four or five miles off, suddenly put up his wheel, squared in his yards and headed down before the wind for us. Our Captain, seeing this, observed to the Mate, "He evidently wants to speak to us. Haul up the mainsail, and haul aback our mainyard." This was done, and while lying there waiting for him to run down, various reasons were given for his wanting to speak us. We were hove to on the starboard tack, and as he approached Captain Sullivan could be seen sitting in the starboard boat with his speaking trumpet in his hand. As he got within hail he raised it to his lips and bellowed out to our Captain a hearty "Good morning." "The same to you," replied our Old Man. This was followed by Captain Sullivan saying, "I have made up my mind to go South. I will haul to a few miles under your lee and you can get my share of the oil ready, run down, drop it overboard and I will pick it up." "All right," was our Captain's answer.

He ran down to the leeward four or five miles and came to the wind on the same tack as us. After bracing forward the mainyard and boarding the main tack, we hauled out casks enough to make thirty barrels from the side where they had been lashed, and put beckets on them as described in preparing water casks to raft. We gave the hoops a good driving, and, with a raft rope lying handy, all was ready by dinnertime.

As soon as all hands had dinner, the wheel was put hard up and we ran down to within a short distance of the *Mitchell,* who had hove aback on seeing us keep off. Everything being ready, the five or six casks were soon tumbled out of the gangway and rafted alongside. Two boats came from the *Mitchell* and caught the raft as it floated out astern.

Our Captain took a boat and went on board her, stopping long enough to have returned to us the raft rope and as many empty casks as we had sent full ones. As soon as the Captain left her, she put up her wheel and headed off to the South. By dark her topgallant sails were dropping below the line of the horizon, leaving us alone on the waste of water to continue our weary cruise.

For a week or two more we cruised in vain; not a whale could we find. The wind had been getting less and less, and now one afternoon it gave out altogether. The ocean was like liquid glass. The heave of it was so slight that the ship would hardly raise or drop the upper course of copper on her sides above or below the water, and no noise was made by the sea on the side except the tinkling of the bubbles coming from the nail holes in the copper as the water now and then reached them. The sun poured down its bright rays with such intense heat and the sides of the ship became so hot that the pitch in the seams swelled from them in small bubbles, melting and running down the sides.

The decks were kept fairly cool by keeping them wet. The poor hogs were fighting and squealing for shady places, and grunting with satisfaction when water was thrown over them while wetting deck.

Below, in the forecastle, the heat was almost unbearable. The watch below lay in their berths trying to sleep, the perspiration streaming from their bodies, with nothing but the curtains drawn in front of their bunks for covering. The cabin and steerage were not so uncomfortable but they were hot enough, even with the big windows in the stern, and all doors between, wide open.

The men on deck at work on various jobs sought, when possible, the little shade afforded by the masts or sails. But the men at the mastheads had to take it straight up and down. The mirrorlike ocean, reflecting the sun's hot rays, made it hard on the eyes; and this, added to the rest of it, made them glad to hear the bell strike for relieving mastheads and wheel. When the sun went below the line of the horizon, a sigh of relief went up from all hands; and shortly after merry laughter was heard from knots of the boys around the windlass, and seated in different parts of the ship forward, smoking and spinning yarns.

The calm continued through the night. The ship with her tall masts and square yards seemed resting on air, her sails hanging flat up and down; not a reef point seemingly moving, but hanging straight from the grommet where it came through the sails. All through the night the ship had stood apparently still, but one could see by watching the compass that she would sometimes turn completely around; this no doubt being caused by the action of the current, which we knew was setting us to the westward at least twenty or thirty miles every twenty-four hours.

At daylight all hands were called, the mastheads manned and decks scrubbed off. The sun rose like a red-hot ball of fire and looked twice her common size. We knew by the look of it that another hot calm day was before us. Just before we went to breakfast, the boat-steerer at the main sung out, "Land, O!" "Where away?" asked the Captain. The reply was, "On the Port beam." By the compass, it bore due West. "That is Sydenham Island, and if we do not get some wind today, I am afraid we shall be set by the current nearer to it before the day is out than I like to be," said the Old Man as he went below to breakfast.

The 2d Mate joined me at the masthead where I had relieved the other boat-steerer after getting my breakfast with the Captain and the officer, as it is rulable for the boat-steerer who has the masthead relief to do; the officer coming up later on. After he finished his smoke on deck and we were leaning over the royal yard, talking about the land in sight and remarking, among other things in connection with it, how similar most of these islands were to each other: "Yes," he said. "But the natives are not. Did you notice how uneasy the Old Man was at breakfast this morning, and how many times he said how much he wished we could get a little wind?"

I told him I had, but that I thought he, like the rest of us, wanted the wind to come so that we could have some relief from the heat and stand a better chance of seeing whales.

"Not altogether that, but something worse than not seeing whales, or being inconvenienced by a little heat. That is to say, we may lose our ship and our lives also," was his startling remark. "This island, that we are drifting so fast towards, no doubt is one of the worst in the Kingsmill group for murder and outrage on any that may be so unfortunate as to get into their hands. We may be set by the current on to some outstanding point of the reefs that surround it, if we have not wind to steer clear. As to lowering our boats to tow the ship clear, it cannot be done, as the men in the boats would be at the mercy of fifty or a hundred canoes in no time. So if the wind does not spring up and we do strike on any part of it, you make up your mind like a man, before another day's heat troubles you, that after today you will think this to be in comparison Arctic weather to what you are surrounded with. In other words, your whaling days most likely will be over."

My reply to his remark was, "Most likely you and I will still be shipmates, if it comes to that"; and I asked him what ships had suffered by these Devil's babies.

His answer to that was, "A number of ships have been lost around here that none have been saved from to tell how it occurred; but pieces of burnt stuff have been picked up by whalers that may account for some of it. Their plan is, after capturing a ship and killing all hands, to take out from her what they care for, which would be all the tobacco first, knives and pieces of hoop iron, and then set her on fire. When they can do so after getting possession of a ship, they run her on shore, then plunder and burn her. The heavy surf would soon remove from sight all record of the crime. This island is one, with some others, that escaped English convicts have settled on and taught the natives more deviltry than they knew before.

"There was a ship lost somewhere around this island, it was supposed, for the reason the last time she was heard of a ship spoke her just to the eastward of here, and on my last voyage we spoke a ship that had picked up a canoe some distance from this island that had been blown at sea and had only one man alive, left from five natives that started in the canoe from shore. He was almost starved and perishing for want of water. They brought him to all right and kept him as one of the crew; named him after the ship, Starbuck. He soon picked up English enough to understand what was told him and make himself understood.

"Afterwards he told the crew about a ship that had been taken, a short time before he left the island, by two white men and the natives, and the ship burnt, after killing all hands and taking all they wanted from her. He said the ship had her four boats chasing whales a long distance from the ship and the canoes boarded her. As there were only a few men left on deck, the white men killed them while the natives held them down. The captain, who was aloft, was the last one killed, by one of the white men, who found a gun in the cabin and shot him. He fell on deck, his head was cut off, and both head and body were thrown to some hogs running around decks. Then they tried to run the ship on shore but they could not work her. The boats, seeing something was wrong, started to come on board. Two boats were stove alongside by throwing pieces of iron and grindstones into them, and the men were killed in the water. The other two boats' crews were killed by the natives in the canoes that surrounded them when they attempted to pull away from the ship. He said that he never helped kill anyone but he got some of the tobacco."

I felt the hair on my head stiffen, listening to the tale he told about this infernal place that was now plain in sight, and towards which our ship was silently but surely drifting. Turning my head in different directions, I hoped to see a light cat's paw of wind somewhere that would give a little steerage way on the ship. Only a few miles either way, north or south, would clear us of this horrible octopus that seemed drawing us to our fate.

The dead calm was in the sky as well as on the ocean. The few light-colored clouds that could be seen were like puffs of cotton, hanging here and there at great heights, without motion. Nothing that gave signs of a breeze could be seen anywhere. The sun poured down its heat equal to yesterday. Turning to him, I said:

"There seems no chance of much wind today. Maybe if we do strike, the beggars, having the ship on the reef, will not murder us without we show fight, which would seem to me useless, as, if we beat them off, where could we go? The only way would be to take the boats and try to get to one of the other islands where they are not so hostile."

"If the ship strikes, no doubt, if we do not show fight, which would be foolish to do, as two or three thousands against thirty or forty of us would be too great odds, they will keep us, after plundering the ship, until some ship takes us off and ransoms us with a box or so of tobacco," was his reply.

"I think we have time, before our masthead is out, to tell you about how some of the devils got served in trying to capture the ship *Triton,* of New Bedford," he said.

"The *Triton* was commanded by a young man of my acquaintance. It was his first voyage as master. Just before sailing, he married a young lady who was called the Belle of New Bedford. After taking a cruise on New Zealand, he stood to the North for the Japan whaling ground, and on his way there he thought to take a short cruise along the Line, through this same group we are now amongst. Not knowing the nature of these natives on this island, he landed in his boat and the natives treated him and his boat's crew to the best they had; took them up to their houses and gave them everything they asked for. The sailors thought it was a Turk's paradise. There was a white man living on that part of the island, but he kept out of sight until later on.

"The captain and men had a good time until the afternoon, and then he told the boat-steerer that came on shore with them to find the men, and get the boat ready to shove off to go on board. The ship had been lying off and on all day. The boat-steerer hunted up the men. It did not take long to do so. They had not gone far from the landing, as the natives kept them well supplied with everything. But when all were assembled, ready to start towards the boat, without a word or action to warn them, the natives pounced on them, tied them hands and feet, tumbled all into a house and put a guard over them. When this was done the white man appeared and told them they had best keep still; if not, every one would be knocked on the head; that he was going to take the boat and go on board the ship and take her, and if successful he would have all hands killed, including those on shore. If not, and the ship got away, then they would be used well until some ship came in sight of the island, when one man would be sent on board of her with a letter stating how much tobacco must be paid for their ransom. With this pleasant information given, the bloodthirsty wretch left them.

"This devil in human shape waited until about sunset, launched the boat, and took twelve or fifteen of the best fighting natives with him, and shoved off. When he left the reef it was light enough for the mate on board the ship to see the boat leave the shore, but before the boat would reach the ship it would be quite dark. This was what he wanted, for it might arouse the mate's suspicions if he saw so many natives in the boat before he got alongside. All worked as he planned until he got alongside. Then, much to the mate's astonishment, he told him the captain had come to the conclusion to send him on board to stay all night, and to tell him to have the ship kept well in with land during the night.

"The mate did not like the look of things and wondered why the captain had not sent him some note. This was accounted for by the pirate saying there was no paper or pencil to write with. After the watch was set, the mate had some talk with the man who had come on board in such a strange way, but could see nothing wrong in his conversation. He seemed perfectly at ease and told the mate that when he left the shore the captain and men were having a splendid time.

"When it came time to turn in, this man said, in offer of a bunk to sleep in by the mate, he preferred to lay on deck with a jacket under his head for a pillow. He would be all right. A coat was given him and he lay down on deck, close to the man at the wheel, with five or six natives amongst the crew and boat-steerers about the decks. Some were laying down among the crew who were having some fun trying to get the natives to pronounce English words.

"Before the mate turned in he left orders to keep the ship pretty well in with the land, and to call him at twelve o'clock. It was about nine o'clock when the mate went below. The 2d mate had the watch, the 3d mate had taken a pillow and laid down in the stern of the waist boat, only telling the boat-steerer who was to call him where to find him when his watch came on deck.

"By ten o'clock everything was quiet. The officer of the deck and the pirate, who said he did not feel sleepy, were talking together just forward the main riggin', with elbows resting on the mainrail on the starboard side. At a signal given by this fiend, two or three natives, who had quietly slipped behind them, caught the officer by the body and legs, and before he could utter a word he was tossed over the rail into the sea, and by the time his head got above water he was so far astern his voice would hardly have been heard if he cried for help. At the same time two or three natives caught the man at the wheel by the throat and body. He also made but little noise as he was thrown over the taff-rail into the sea.

"While this tragedy was going on aft, the boat-steerers in the waist had been disposed of, one thrown overboard and the other, knocked senseless, laying stretched on deck. The men on deck had been attacked at the same time. Some had escaped down the fore hatchway, which was open, some run into the forecastle, others had tumbled overboard or been knocked over and lay dead or senseless on deck, the heavy oak heads of scrub brooms with handles in them four or five feet long affording murderous weapons in the hands of these bloodthirsty wretches—these laying around the try-works right at the time, ready for them to use.

"The pirate and his natives now had possession of the decks. All who were not killed were below, except the 3d mate, asleep in the waist boat, who had been awakened just about the time when the last man was tumbled over the rail into the sea. He was not seen by the natives as he sat in the boat, with only his head above the gunwale, but he could see by looking under the foot of the mainsail which left a space of a foot or so between the boat's gunwale and the bottom of the sail—the ship being on the starboard tack with the main sheet hauled aft—between the stern of the boat he was in and the head of the larboard boat.

"He was almost paralyzed with horror when he saw the natives had charge of the ship. He thought of cutting the falls to the tackles that held the two ends of the boat, and the grips that held her to the side of the ship, and letting her roll off the cranes into the sea, taking a slim chance that way for his life. But, thinking before he could cut her clear and roll her off the cranes he would be seen and killed, he abandoned the idea and crept forward in the boat, showing as little as he possibly could of his body above the gunwales, to where the lances were tied to the boat's side. Quietly casting off one of these, he crouched in the boat with this in hand, the sharp head pointing inboard, resolved to sell his life as dearly as possible. He looked hard to catch sight of the bloodthirsty white man, thinking, if he could get a chance to put the lance through him before he was killed himself, that he would die with less feelings of regret.

"He had not been watching long before he saw him crouching along the deck, working his way towards the boat he was in, peering towards the stern, thinking perhaps he was in some of the boats as he could not find him in his stateroom below where he had hunted for him after killing the mate, who was asleep in his berth when this fiend had split his head open with an ax.

"The 3d mate held his lance ready, which owing to the darkness this wretch did not see, its wicked head sticking through between the lanyards of the mainriggin', or any part of him who held it, as he now had dropped his head below the gunwale of the boat, ready to spring up when the other approached near enough. Now judging his time right, the 3d mate raised on his knees. At the same time, the other man sprung towards the boat, brandishing in his right hand a fleshing knife." (Such as is used to cut off the pieces of meat that sometimes adhere to the blubber when cutting in a whale; the blade of which is about eighteen inches long.) "He made his last spring, for the 3d mate darted his lance with such force and skill that the head of it went through the body and into the deck, pinning him fast on the four or five feet of its shank.

"His yells were fearful, as he in his wild efforts to clear himself tumbled around the lance that held him to the deck as a butterfly would be held to a board with a pin through its body. The natives seeing him in such a queer position, with the blood flying from him and he fast to the deck, and not knowing how it occurred, got panic-stricken. Instead of helping him free himself, they hid themselves under the windlass and around the tryworks, leaving him in the hands of the Devil, as they thought he must have done the act. Their superstitious feelings gave the officer a good chance to crawl over the stern of the boat behind the lee clew of the mainsail on to the ship's rail, out of their sight, and in on deck. Keeping close to the rail on the port side until abreast of the cabin companionway, then making a sudden dart, he was quickly down the cabin stairs.

"Hoping to find some of those below alive to help him take the ship back, he looked into the mate's stateroom. The ghastly sight of the dead mate's body, half out of his berth, told him no help could be had from him. Next stateroom was the 2d mate's. Of course no one was there. His stateroom was occupied by himself and the steward. The steward was not there. He sprung to the door that led into the steerage from the cabin. Opening it, he found all in darkness and not a sound to be heard; no one was in the berths. This made his heart almost stop beating from the thought that no one was left of the afterguard but himself, but, as he had seen no one but the mate dead in the cabin, or so far in the steerage, he thought perhaps the steward, cooper and some of the boat-steerers might have crawled on top of the casks in the afterhatch between decks.

"The door leading from the steerage into it was open. He stepped quickly through it up to the tier of casks that stood just forward of the afterhatch, and sung out, 'I am the 3d mate. If anyone is stowed away in here, come out as quickly as you can, and help me take the ship back from the natives, who have killed everybody on deck and the mate. Do not be afraid, but come out quick, before the natives get over their fright of seeing that head devil of a white man pinned to the deck by a lance through his body. That is him you can hear now, yelling, but he don't yell so loud or as fast as he did. You need not fear anything from him, so come out quick.'

"Hearing the 3d mate's voice, those that had stowed themselves away on hearing the groans of the mate when this pirate was killing him came out from their hiding places amongst the casks, and they went into the cabin to gather up what weapons they could lay hands on. There were only four of them to retake the ship from twelve or fifteen natives. The 3d mate took from under his mattress the two boarding knives that he had charge of, and that are never left but in some safe place, as they are dangerous tools to have in a place where a desperate man might get hold of them. The blades are about three feet long, sharp as razors, having long handles, turned round to fit the grasp of both hands. The blades are two-edged where they fit into the handles, have a width of two inches running to a sharp point at the end, and are thick in the middle to keep them stiff. A very light push on one placed against a man's body would send it through as easy as a fork would go through a dish of cold mush. Placing one each in the hands of the cooper and boat-steerer, he asked the steward where the pistol and sword that belonged to the captain could be found. He soon had hold of them. Finding the revolver, which was a Navy Colt, all loaded ready for use, he slipped some cartridges loose in his pocket.

"Giving the sword to the steward, he gave them the plan of action. They would creep up the cabin stairs, ready at the word he would give when he saw fit to make the rush forward, two on the port and two on the starboard side, not stopping for anything until they reached the forecastle scuttle, and reaching there to get the men on deck as soon as possible if any were alive. If none were left, then to sell their own lives as dear as possible. The chances were against them, but they might still get the ship back as the natives had lost their leader. To encourage them, he said, 'No doubt but what there are eight or ten men cooped up in the forecastle and, getting them out to help, we can retake the ship in half an hour.'

"He cautiously crawled up the cabin stairs, the others following close after. Leaning over the top step with just his eyes past the side of the companionway, his head close to the deck, he could see the deck forward without being seen himself. After he had taken a look on both sides, and aft around the wheel, he in a whisper told them no one was aft and no one in sight on the starboard side, but on the port side he could see five or six natives coming towards the fellow that was pinned to the deck, who between his curses seemed to be telling them to do something. 'Now,' he says, 'we must move quick. That devil of a white man seems to have life enough to raise hell yet if he gets those natives to pull that lance out of the deck and clear of his body. The steward alone must take the starboard side of the deck. I do not think he will meet anyone on that side. If he does he must use his sword and get to the forecastle as soon as he can, tell the men to get on deck, arm themselves with iron belaying pins, sticks of wood, or anything, and rush to where they see any fighting going on. Us three must take the port side and get between that white man and the natives.'

"By this time they were all on deck, crouching behind the companionway out of sight from anyone looking aft from forward. The natives, half panic-struck, creeping towards the howling and cursing wretch, did not notice in the darkness of the decks the three men advancing with bare feet rapidly towards them, until the boarding knives and pistol were doing murderous work amongst them.

"One thrust of the boarding knife through their naked bodies soon opened the way, and caused those that could to run in every direction. The steward reached the forecastle without trouble, and the eight or ten men there had now got on deck, arming themselves with whatever they could, and joined in the chase of the naked devils. In a short time there was not a live one left on deck. Some had jumped overboard and no doubt were swimming for hours, until they drowned or got eaten up by some shark.

"The natives being cleaned out, the white man was looked after. There was not much life left in him for he had been thrust through with a boarding knife two or three times during the first part of the recapture. It took two men to pull the head of the lance out of the deck, they having a good hold of the pole in doing so. No doubt a shark had him eaten before he could have been in the water half an hour."

I had been listening to this story of the 2d Mate's with such eagerness that I hardly had paid any attention to looking out for whales or anything else, and when he suddenly broke off his conversation and said, "Damn them, here they come!" I almost jumped off the crosstrees on which I was standing.

Looking in the direction of the island, which was now plain in sight, there could be seen numbers of black spots on the mirrorlike surface of the ocean. I spoke up, saying, "There must be a lot of canoes if all the black spots I see are canoes." "Take the glass and look at them. It is likely you will see with it twice as many more, and as we get nearer to the land more yet will be seen," was his reply. On looking at them I could see at least fifty or sixty, all headed for the ship. Putting up the glass I said to him, "How many men do you suppose are in each canoe?" "Well," he replied, "if they have no women in with them there would be perhaps four to six in each one, but they no doubt have some women; just enough, likely, to make it appear that they do not mean bad, and to get a deck hold. But the Old Man won't allow that if we can help it."

I was not feeling very hilarious, but still wanted to know how the affair terminated with the ship after they had retaken her, so I asked him about it.

"Well," he said, "there is not much more to it. After they tumbled the dead natives overboard, laid those of their own crew who had been murdered out for burial, and washed off decks, the ship had lost, with the captain and his boat's crew, fourteen men. The ship, when daylight broke, was out of sight of the island, but they thought it no use to try to work her back to it. As they thought the captain and those men who went on shore with him had been murdered, the best course to pursue would be to take the ship to the Sandwich Islands and turn her over to the Consul. The men who lay dead were given a proper burial. When this was done, sail was made on the ship and a course laid for Honolulu, to which place she safely arrived in a few weeks' time. The Consul took charge of the ship for the owners, sending word to them by the way of San Francisco and an account of the affair to Washington.

"The funny part of the whole thing was: after the ship had been laying in Honolulu some time, the Consul waiting for instructions to know what to do with her, a ship came in one day that had on board of it the captain and his whole boat's crew. On coming ashore the captain reported to the Consul he had been treated well by the natives after his ship left him, and had been released by the captain of the ship he came on, paying a ransom of two or three boxes of tobacco. He also said not one of the natives that went on board the *Triton* had been heard of after, and he had felt sure the ship had not been taken; was sorry so many lives were lost, but glad to get back his ship again. He shipped more men and officers and went on with his voyage. The 3d mate got a good deal of credit for his bravery."

By the time he had finished his yarn, the bell struck for the relief of the mastheads. The 4th Mate relieved the 2d Officer, a boat-steerer took my place, and we went down the riggin' on deck.

By the time we arrived on deck the canoes could be seen from the rails. The island was about six or eight miles off and we seemed drawing towards it faster than ever. The Captain had the steward and boy bring on deck what firearms were in the cabin, consisting of some flintlock muskets, a few revolvers, and one double-barreled shotgun. These were loaded and stood against the mizzinmast. The lances were taken out of the boats and placed on top of the tryworks, ready for use. The cutting spades were taken from the racks overhead, where they were kept, and laid along the decks, ready to be caught up at a moment's notice. All ropes that were outside the ship and could be caught hold of to help anyone climb up the side were hauled in.

During the night I had had the toothache. It was still causing me a great deal of pain, so I went to the Mate and asked him to pull it out. He was always willing to do anything in the dentist or surgeon's line for anyone on board the ship. He seemed to take pleasure in cutting or hacking on the human frame. He ought to have been a surgeon. When I spoke to him of what I wanted done, he stopped in what he was doing, looking at me for a minute straight in the eye, and suddenly broke out with the words, "I will be damned if I ever heard of such a damn fool thing in all my life. Here you are with three or four hundred black-skinned, whooping, roaring ugly devils just ready to board us, and if they do they will most likely feed the sharks with us before the sun sets—and you want me to p-u-l-l out a t-o-o-t-h. Great Guns and bags of gold, who would have thought it? What in the name of Old Moll Row, or any other woman, do you want it done for just at this time, even if it does ache?"

"Well, I will tell you," I said. "If the ship is taken by those fellows, some of us may be saved, and among those I may be one. Now, the 2d Mate tells me that about the only thing they have to eat is coconuts. I shall starve to death if my jaws are not in working order."

He stood looking steadily at me without saying a word for a second or two, wheeled suddenly around, and started for the cabin, muttering. "Coconuts, be damned. I will pull your head off if you want me to."

It was not long before he came out of the cabin with a pair of those old-fashioned tooth pullers in his hand. Sitting me down on the after corner of the main hatch with my head leaning back against a coil of riggin' hanging from a pin in the fife rail around the mainmast, he looked into my mouth and found the right tooth, shoved the instrument of torture over it, and bringing a sudden jerk on it, brought it out of my mouth with the tooth in its claw. Thinking by the shock he gave me he must have taken the jaw with it, as I wanted nothing removed but the tooth, I remarked as soon as possible, that if he had no further use for my jaw to please put it back where he had taken it from. "Pooh," he said, "you will not miss the piece I have. It is not more than half an inch long." He was right, I did not miss it, or some three or four more pieces that afterwards came out from the place he had fractured by letting the claw catch below the tooth, I suppose. "Such times as these, people must not be too exacting," he said. I thanked him.

It was about time, now, to make preparation for keeping the blueskins from boarding the ship. The Captain told the Mate to call the men down from mastheads and hoist all boats up to the davit heads, so the men would have no trouble handling the spades and lances under the boats against any that might attempt to crawl up the side under them; also, to station the men at working distance apart from bows to taffrail, each one armed with a spade or lance. The Captain would take charge of the quarter-deck, 4th Mate on one side, he on the other; 2d and 3d Mates amidships with three boat-steerers; and the Mate with one boat-steerer forward.

All was according to orders, and when the first canoes came close alongside and attempted to make fast to the ship, over the side came the bright gleaming spades and lances, with their sharp edges close to their hands reaching out to make fast on anything to hold the canoes, which caused them to shove off with shouts of fear and eyes sticking out like crabs'. After the first lot had tried it, they paddled out a good ship's length on each side of us, forming a line parallel with the ship that extended a short distance ahead and astern and was added to by canoes constantly arriving for an hour or two, until they formed a cluster twice the length of the ship and at least five or six deep. They were on both sides, jabbering and gesticulating with a din and uproar that made things hum. There must have been at least five hundred persons. Out of the number perhaps there might have been fifty women. Now and then some canoe with women in it would try to come alongside, but a clip with the sharp spade that took off from its side a sliver would send them paddling furiously away again.

The Captain gave orders not to kill any if it could be helped, unless some white man should be in one of the canoes and attempted to board the ship; and if he should attempt it a second time after being once warned, to kill him on sight. Now and then one would shake his fist at us. Sometimes one would brandish a short sawlike sword made of a strip of coconut wood, having on each edge the teeth of a shark lashed firmly the length of it and leaving just enough to form a handle—a wicked thing to use on a man's head or face, as, when they strike with it, at the same time it is drawn back, cutting the flesh like a saw from half to an inch in depth.

The Captain, standing on the poop deck with his musket in hand, seemed to afford them a good mark for ridicule. They jeered at him, made faces and signs of cutting his head off, and pointed to the island, as much as to say, "When you strike there, off will come your head." The old fellow stood it well. He would walk back and forth, whistling for a puff of wind to help carry us out of our trouble, and mutter an oath now and then.

This had gone on for some time, and we had got within three-quarters of a mile of the reef and midway of the island. The ship, which up to this time had been drifting straight as though steered for it, could now be seen to draw by the land almost as fast as she neared it. The Captain sung out to the Mate, "If the reef under water does not bring us up, there is a good chance of our getting clear all right, after all." "Yes," said the Mate, "we seem to draw by faster now than we drift on the land. If we could put two boats ahead we could soon swing her clear, but the men in the boats would have no chance against the large numbers of natives that might pounce on them, so that cannot be thought of."

The natives about this time showed by their actions that they thought we might clear the island, after all, and they would lose the chance to pick the old ship's bones. It made some of the men in the canoes very demonstrative. Five or six of the canoes that had no women in them started from the crowd, paddling rapidly towards the ship. Each canoe had five men in it. As about the same number were approaching from the opposite side, it began to look like business had opened up; but we did not fear them much, as they had no firearms, and the shark's-teeth swords and spears they had would be dangerous only in close quarters, where we who manned the side would see that they did not get.

The Captain saw the movement and sung out, "Look out there for those fellows! Do not allow any to stop alongside but do not kill any if you can avoid it."

What was their motive in rushing in on us that way, I failed to understand, for when we put out the spades and lances against them as the canoes came dashing alongside, they sheared off and paddled away as fast as they had come, back to the crowd of canoes again, yelling and shouting like blue lightning. Now and then one canoe would paddle halfway out from the others towards the ship, and someone would raise up and yell out a lot of lingo, at the same time gesticulating rapidly, which would be answered with loud shouts accompanied by the pounding of the paddles in the other canoes in the water and against the sides. One or two of the orators, after winding up their harangue, would with the most indecent insults take their seat in the canoe, and away they would paddle.

By this time the Old Man was overrunning with rage. He sung out for someone to pass him up the shotgun that was loaded with double-B shot. Taking this in place of the musket, he swore to make it warm for the next fellow that attempted that insult again.

It was not long after that, before a canoe shoved out again and commenced the same harangue. This time the orator was a large dignified-looking chap, who had a big white shell fast to a string around his neck. His motions were all according to rule, and when he turned around in the canoe to go through the final act, as the others had done, Chesterfield could have been no more precise about it than he. To give his bow, backwards, more effect, he had placed a hand on each side of his person. As he bowed very low to give it all the effect possible, no clothes obstructed the shining mark. The Captain raised the gun to his shoulder, taking sure aim at the bull's eye, and pulled the trigger. The next instant the native, with the same dignity as in all his former motions, went headfirst into the water and that was the last seen of him.

No doubt he swum under water to the outside canoes and crawled into one, as quite a stir could be seen shortly afterwards amongst some of them. The canoe the dignitary had disappeared from lay quiet for a few minutes. The men in it seemed dazed, looked towards the hole in the water made by the sudden dive of their leader, then dipped their paddles into the water and furiously paddled the canoe into the center of the others. Not a word was heard from any of the natives for a time. Then a shout went up from them that almost made the reef points, hanging up and down against the lifeless topsails, go pitter-patter.

"Look out now!" exclaimed the Captain. "Some of the bold ones may make a rush. If they do try to board us, keep them off, but do not kill any unless you cannot help it."

In a few minutes a commotion could be seen amongst

the canoes. Some six or eight on each side of the ship separated out from the others. In each one of these were six men. They seemed stout, ugly-looking fellows. The canoes took positions on both bows, beams, and quarters. After a few words with each other they commenced paddling towards the ship, coming alongside about together. Four or five in each canoe dropped their paddles like so many fool monkeys, caught hold of the chain plates and moldings with their fingers, and tried to climb up the ship's sides, which were bristling with steel fore and aft. Confusion reigned among them a few minutes after, as they tumbled back into the canoes and overboard, many of them bleeding from the cuts received on their bodies, arms and heads. None were killed outright but some could be seen hanging partly over in the canoes as they paddled slowly away, using but half the number of paddles. These canoes made straight for shore.

The body of those canoes that the boarders came from had drawn close in around the ship from both quarters, forming a circle around the bows; and of what must have been their reason for doing so, we could form but one idea. That was, in case the first lot had been able to get on deck, the others would rush in and overpower us, not taking into consideration that it was impossible for them ever to get through the line of glistening steel that guarded the ship's side.

No more attempts at boarding were made by them, nor were any more demonstrations of insult offered of the kind that had afforded a target for the Captain's workmanship. Both attempts to board having failed, they seemed satisfied that it was no use to think of getting charge of the ship that way.

We had now drifted in close enough to see the bottom under the ship and had neared the reef so that it was less than a quarter of a mile off. The critical moment was on us.

Fifteen or twenty minutes now would tell the story, as the ship was very near the turning point of the island.

Five minutes passed. A dark-looking mass of coral passed slowly by on the starboard side, the copper on the bottom clear of it but scraping it just enough to break off some fringelike prongs, but touching it so lightly no jar was felt in the ship. At the same time other patches could be seen here and there on each side of us.

The natives now commenced to shout in a most infernal manner, rising up in their canoes, tossing up their paddles in the air, catching them by the handles when they came down and swinging them around like war clubs. No mistaking, the motions meant they would soon be beating out our brains, as the ship would soon strike and we would be at their mercy.

Ten minutes more drew out its long extent, the ship still drifting clear. The strain amongst us was easing up a bit for it now seemed that the distance had certainly widened between the ship and the reef out of water, on which the breakers were lazily showing a roll of milk-white foam where the blue edge of the sea crumbled itself.

The suspense for the last half-hour had been so great that no loud word had been spoken. The Old Man walked the poop deck with nervous steps, now and then gazing over the side down into the water, then towards the shore on which crowds of men, women and children could be seen among the coconut trees and around houses built of grass that lined the shore, back of a white coral-sand beach.

The Mate in the bows suddenly caused every man's heart to jump into his throat by singing out to the Captain and pointing at the same time ahead: "Here is a patch of coral right across our bows just under water. The ship can never get over it."

"How far are we from it?" asked the Captain.

"Only about three ship's lengths," replied the Mate.

"How much on each bow?" was the next question.

To which the Mate answered, "On the starboard bow I can see no end, on the port bow about four points is the outer end."

This was apparently our final resting place, as from what he reported the ship would likely bring up on it in a few minutes. The natives thought so too, for those on shore could now be heard joining their caterwauling to those in the canoes, and they were dancing up and down with delight. But we had hardly made up our minds that there was no show for the ship when we all took notice that she commenced to turn around and to drift almost at right angles seaward from her former course; and in less than ten minutes we swept by the obstruction, all clear by fifteen or twenty feet. In half an hour we were in water a mile deep, by the look of it.

When we saw all danger past, did we not yell in derision to those blue-bellied beggars, who had stopped their clatter on seeing the ship pass what they made sure would be her doom! The guns and pistols were all fired off. At the same time three cheers were given, as they turned and paddled ashore.

By sunset the island was hull down, the decks were all cleaned up, and the spades and lances put away in the places where they belonged. The calm continued through the night. A light wind from the S.E. came up with the sun. With all sails swelling out from yards and clews, there was the merry rippling swash of the water under the bows as the old *Charles W. Morgan* slid through it without hardly raising or dropping the end of her fly jibboom. Adding to this, the coolness of the fresh morning breeze sent such feelings of relief and pleasure through the crew that they acted more like a lot of wild schoolboys than men who had been almost baked by the sun's heat two or three days and had just got clear of a shipwreck, if nothing worse.

The wind increased during the next twenty-four hours until it blew a good topgallant breeze. It was quite a change for a few days from the light winds we had been having; then we had light winds again for the rest of our cruise, but no more such calms as came near piling us up on the reef of Sydenham Island.

We felt more thankful than ever about our escape from going ashore on Sydenham Island, after speaking the ship *Two Brothers* and hearing her report of a ship belonging to Sydney which had been lost on that island only two or three weeks before our escape.

It seems that this ship must have run on the same reef under water that we passed in getting out of our scrape. She, however, went ashore in the night, during a squall of wind and rain. She was a whaler and the captain had his wife on board. When daylight broke he took his wife in one of the boats and pulled on shore. On landing, the captain and men were immediately knocked on the head and the wife was carried away as a slave, which is the custom of these devils with all women captured by them. Another boat that was following the captain on shore was surrounded by natives in canoes, and the crew killed. Those on the ship, seeing this, and large numbers of canoes coming off to the ship, immediately lowered the other two boats. Throwing into them what few provisions could be got handily, and also a little water, they shoved the boats off from the ship, set the sails and stood out to sea, preferring to take the chance of being picked up by some ship before they starved than the sure chance of being killed by attempting to land. They were picked up a few days after by a ship the *Two Brothers* had spoken.

There was no sign of the ship when we went by the island. She had been burnt, no doubt, and the unburnt portion had most likely slipped off the reef into deep water. Months after, it was found out the captain's wife had met a fate far harder than that of those who had been killed outright. If this occurrence had been known to our Captain and officers at the time of our encounter with the natives, there would have been some less of their number left to perform such acts of outrage, as, let our fate be what it might, never from the start would one have escaped alive that could have been reached with anything to have killed him. This feeling was expressed by all, from the Old Man to the little cabin boy.

Strong's Island

WE CONTINUED our cruise for some six weeks longer and took whales enough to make us about two hundred and fifty barrels. When we had got into the Long. of 165° E., Lat. of 7° N., the time was drawing near for turning our jibboom to the South for a cruise of the New Zealand whale grounds, but, as we needed wood and water, the Captain made up his mind to make a port of an island about one day's sail from the position we then occupied, called by sailors Strong's Island.

The course being laid for it N.N.W., all sail was set, with fore-topmast and main-topgallant studdingsails sent out. In twenty-four hours' time, with a fair fresh breeze, the island hove in sight. We did not get near enough to anchor on the day we sighted it, so at sunset the studdingsails were hauled down and the ship brought to the wind under whole topsails. During the night we laid off and on. At daylight, the island being some eight or ten miles off, all sail was set on the ship except the studdingsails, and she headed towards it. Shortly after breakfast we had approached within two miles of it and luffed the ship by the wind with the main-topsail to the mast.

The Captain had never been at this island before, nor had any of the officers, and no chart we had on board gave any directions how to find anchorage. The Captain started in a boat towards the land, telling the Mate to keep the ship close in, and, upon seeing the boat with a flag set, to keep the ship for it, as he would have found the anchorage.

The boat pulled in to the land and I, being at the mast-head, could see her go out of sight into a line of breakers. Knowing the Captain must have found a passage through a reef that seemed from where I stood to form a barrier from two points across a deep bay, I reported to the Mate the fact of the boat's disappearance through a passage in the reef. "All right," he replied, and shortly after gave the order, "Wear ship," as the boat had entered the reef on the ship's lee quarter. Wearing around until the ship pointed towards where I last saw the boat, the wheel was steadied and the yards trimmed. After running by that course until we got within a short half-mile, the yards were braced sharp up and the ship brought to the wind. The Mate hailed me and asked if I could see anything that looked like a passage through the reef.

I told him that I could see a passage into a small beautiful bay, but hardly thought a ship could go through it as it was very narrow, although it seemed deep and short.

The ship's head was then brought to the wind and the yards braced sharp up, and we tacked back and forth for a couple of hours, until the boat with a flag set could be seen pulling out of the passage.

"That means the place," said the Mate. "Hard up the wheel!"

"Aye, aye, sir!" said the man at it.

"Square in the after yards," was his next order.

We headed now for the boat and soon ran down to it.

The Captain, who was in the boat, called out to the Mate, when near enough to hail the ship, to keep right ahead on the course now steering; that he could get out of the boat without luffing the ship by the wind. The wind being light, the boat swung alongside without any trouble and the Captain and boat's crew came on deck, leaving one man in the boat to keep it off from the ship's side.

Arriving on deck, the Old Man told the Mate that it was the prettiest little harbor he had seen for many a day, although the entrance to it, through the reef, was very narrow. "You can see the opening now, in the line of breakers, just as the jibboom points," he said. "Have you the anchors all ready to let go?"

"Yes, sir," replied the Mate.

The passage could now be plainly seen no great distance off, showing a smooth opening in the milk-white, dashing, roaring breakers that were sending up sheets of foam as the sea rolled against the obstruction on each side of it. The passage was perhaps one hundred feet in width and very deep, so that the sea had no chance to form a comber but had to be contented in raising a slight roller at the mouth, which soon ran down in the smoothness of the lakelike harbor, completely land-locked.

It looked queer to see the ship heading for that little opening not as wide as her length, where a sudden whirl of a few spokes of the wheel, either to port or starboard, would send the noble ship with her lofty masts and white swelling sails to destruction in a few minutes. By looking on either side, as she sent her jibboom into the passage, sharp-pointed masses of coral rocks could be seen, against which the sea exhausted its force and settled back with loud and angry roars that might mean, "We failed to tear you asunder this time—but look out, when we roll in on you next time!"

Hardly had we entered the passage when came the orders from the Captain, who stood on the bows between the knightheads, conning the ship: "Haul up the foresail! Haul down the jibs! Clew up topgallantsails!"

These orders were quickly obeyed and the ship now was running under her three topsails. As we had been running before the wind, the mainsail had been hanging in the buntlines.

From the entrance to where we were at the time sail was taken off the ship might be one thousand feet. Then we shot into a bay of glassy smooth water of an oval shape, the passage to it about one-third its length from shore, where it ended on the left. To the right could be seen a longer stretch of water, extending some way past the point of land that joined the reef we had passed through. We had hardly time to glance at the surroundings when the Captain gave the order to: "Port the wheel!" Then, "Brace up the topsails, port braces!" And, "Steady!" as the ship luffed and headed towards the head of the lovely lakelike harbor, shooting past the point of land on our right.

Two or three ship's lengths later came the next order: "Clew up the topsails! Stand by the anchor!"

The topsail halyards were let go by the run, at the same time the topsail sheets were cast off from the pins that held them, and before the yards had hardly settled on their lifts, all three topsails were hanging to the yards by clewlines and buntlines.

"Hard a port your wheel!" The ship, having still way enough, turned her head towards the right-hand shore, which was now some three ship's lengths off. "Let go the anchor!"

"Aye, aye, sir," replied the Mate, at the same time saying to the boat-steerer who had it all ready, "Let go ring stopper! Stand clear the chain!" Down went the anchor into the water with a splash, and the chain flew out the hawser hole and around the windlass with a rattle that woke the echoes from the highlands on three sides of us. After giving the ship thirty fathoms of chain on her anchor, which lay in four fathoms of water, the men lay aloft and soon had all sails furled.

After the riggin' had been coiled up and the decks swept off, we took a look at our surroundings. The ship lay within three hundred feet of the beach that was composed of sand coral and disintegrated lava. This beach extended to our right until, joining the reef we had passed through in getting in, it followed around a point out of sight from where we lay. On our left, looking towards the land near us in both instances, the beach followed the land until it was lost around a point that was near the main island, beside which was a passage to the sea for canoes. By this we found the harbor formed from an island that had two points running into the sea about two miles apart, the widest part of the bay joining the shore on the main island about a mile distant. The mangrove trees all along on that side gave the appearance of a thick jungle of shrubbery growing in the open water. No beach or any landing could be seen when the tide was high.

The land on the main island rose quite abruptly to the height, I should think, of twenty-five hundred feet, and was covered thickly with tropical vegetation to the summit. The small island was in the highest place, say, fifty feet, running from that point each way fore and aft (lengthwise) with a regularity in its outline that a batch of dough would have, when poured out of a pan, before it had time to flatten. The island was not much over half a mile across; and the whole affair was volcanic, with trees and shrubs growing in pockets of alluvial and in the crevices amongst the rocks.

Not many houses could be seen from the ship. A large house stood out clear from the coconut trees that stood quite thick just back of the beach, and this was pointed out as belonging to Captain Hussey, who had left his ship and taken up his home here, for reasons I will speak of later on. The strange thing was the few canoes they seemed to have. Up to the time we went to supper, but two or three unseaworthy looking affairs had been alongside, they bringing only a few coconuts and some wild mountain bananas that were better cooked than eaten raw. The natives that came in the canoes seemed such downhearted, broken-down, lifeless things that it caused great surprise to us all; but later on I, for one, could see a good reason for it. Those we saw first were a fair sample of their class. Their color was lighter by a number of shades than those we had seen among the Kingsmill group when cruising there, and their features more regular; but physically there was no comparison.

Shortly after the decks were cleared up, the Captain had gone on shore to visit Captain Hussey, sending the boat back for bread, flour, tea, coffee and other stores, with word to the Mate that he would make his quarters in Captain Hussey's house while the ship remained in that port.

After supper the Mate told the other officers that everybody but a boat's-crew watch could go on shore if they wanted to, as there would be no liberty given during the week days. The 3d and 4th Mates with most of the crew went on shore; but I, having the first watch with my boat's crew, did not join them.

The Mate and the 2d Officer stopped on deck most of the watch, smoking and talking. The night was lovely, with a small moon just large enough to cast a faint shadow of the ship's tall masts and yards on the glassy smooth water; so still that only a faint tinkling sound could be heard from the beach as the ripples tried to creep up and run back on the shore without disturbing the deathlike silence of the secluded and wonderful little place. The stars shone with such a clearness from the dark blue sky that the smallest one that could be seen by the naked eye could be plainly observed. Not a sound from shore could be heard, any more than from a graveyard. The officers and men on deck talked in a low tone of voice, seeming fearful to break the peaceful silence.

Walking aft to where the officers sat, I spoke to them of the holy silence of the night and asked them if it could be possible that such a peaceful place could ever have been disturbed by such acts of horror as many of the South Sea Islands had been.

The Mate spoke up and said, "Beyond a doubt this island, like many others, has had lots of fighting and killing amongst themselves, but of that I have never heard, and seeing so far but few natives, it leads me to suppose that plenty of that going on has reduced the number."

Turning to the 2d Mate, he asked him if he ever heard of that brig being taken here and burnt by the natives? The 2d Mate replied that he had heard something about it, but did not know how it occurred.

"It seems," said the Mate, "a brig that was sailing from Hobart Town, on a whaling voyage some two or three years ago, put in here for wood and water, as we have. She had some liquor on board, which was handed around pretty freely, the captain keeping drunk most of the time. During the time this was going on, quite a number of quarrels had happened between the natives and crew. At last some trouble about women occurred and the natives rose on them one night, when most of them were on shore sleeping around at different houses, and killed them, the women helping. As soon as the men on shore were disposed of, they went on board the brig and killed the captain, who had gone to bed drunk, and slaughtered the rest of the men except the cabin boy. Him they took ashore and he lived some time with them, until a ship called and he escaped on board of her. The natives plundered the brig of what they wanted and then burnt her—the hulk of her lies somewhere on the bottom beneath us."

"Well," said the 2d Mate, "if they wanted to do so, they

could play some such a game on us, if the men are allowed to sleep on shore."

"I don't think there is any danger from them now. The instance I spoke of is the only one I know, and besides, Captain Hussey has been living here amongst them for some time," said the Mate, at the same time rising up from his seat. Saying he was going to turn in, he told me to keep a good lookout that he did not wake up with his throat cut by the beggars coming on board.

With a word or two more about how pleasant the evening was, he and the 2d Mate went into the cabin, leaving me the deserted deck. The five or six men on the deck forward were stretched out full length, with a coil of riggin' for a pillow.

Walking the deck and thinking, I told myself that somewhere under the ship lying on the bottom were the charred bones of men who had gazed on the scenes now around us; men who perhaps had as much affection for their homes and friends as I, and looked forward to meet on their return fond mothers and loving brothers and sisters, cut short of life by the hands of some of these same savages that now were almost within hail of us. Such notions running through my mind caused the night to have a different look, and the moon settling behind the hills seemed to add by the loss of its light to my dismal thoughts. When the boat-steerer and his crew that relieved us came off from shore at ten P.M., I went below and turned in feeling far different than I did at the first part of my watch. I soon fell asleep, though, and had no bad dreams, as I recall.

AT DAYLIGHT all hands were called and by noon we had rafted empty casks enough to hold at least 150 barrels of water. When dinner was over three boats took the raft in tow. A native went in our boat to point out the place we had to go, as the shore line on the main island towards which we were pulling showed no openings in the mangrove branches that were sweeping in the water. After we had pulled almost near enough to send our boat crashing into the branches, the native said something in his lingo that sounded more like telling you to go to hell than anything else, and waved his hand furiously for us to turn the boat to the left. After a few strokes more, an opening could be seen a ship's length wide.

We pulled for a little time longer and came to an opening in a bank of mud through which, by hard pulling, we succeeded in placing the raft. The mangrove trees were now quite a distance away from us on each side, and we were in a stream of pure fresh water fifteen or twenty feet wide, flowing quite swiftly towards the bay and carrying a depth of four or five feet.

Making the ends of the raft rope fast, we knocked out the bungs of the casks, and by rolling them partly over we soon filled them. During this time, eight or ten young girls were tumbling about in the water, close to the men at work; and the most timid man in the crew at work there said he was not afraid, even if they had left their bathing clothes at home.

By dark we had the casks alongside and hoisted aboard. After getting supper, the Mate said the Captain had told him there would be no need of a watch, but that during the night any officer or boat-steerer who happened to be sleeping aboard could take a turn now and then on deck; and a boat with a line fast at each end, one on shore and one fast on ship, could be used by anyone to come or go in, without having to hail the ship or shore. This was rendered easy, as, during the time of our being away filling water, the Mate had run out a hawser to the shore, making it fast to a convenient tree and taking the other end through the hole in the quarter (used for the head rope in cutting a whale) to the capstan. Heaving the same taut, he kept the ship from swinging around her anchor and brought the stern within 100 feet of the shore. This was perfectly safe to do, as but very little wind ever blows there.

From what the other boat-steerers told me about what they had seen on shore the night before, I came to the conclusion to take a turn myself, so, taking as the rest did a lamp that had a tubelike fixture in the bottom that would fit over a stick three or four feet long, I went on shore with two or three others. As it was dark when we landed, we lit up as soon as clear of the beach and into the shrubbery, those who had been on shore before going ahead.

The night was perfectly calm, with no more flicker to the wicks than being in a closed room. After stumbling over a path of rough stones with thick bushes on either side, we entered what seemed to me a deep cut, the sides some twenty to thirty feet apart and at least twenty high. Of that I could not determine, as the lamps would not throw out light high enough to see, but the stars could be seen overhead where the trees and vines left openings.

"Hold on, boys," I sung out. "What in the name of thunder have we here on each side of us?"

The reply to my question was, "Stone walls." Sure enough, they were, and, my sight getting better as we passed along, I could see many enormous rocks in these walls. After stumbling along between them for a time, we came to an opening in the one on our right. The sides of this opening were about ten or fifteen feet apart, and it could be seen that the large flat stones were laid with such care and precision that a stone mason might be proud of the work.

The boys ahead turned into this open gateway and I followed. After entering, we found a large open space, covered with stunted grass, in front of a long thatched house that could have been no less than one hundred feet in length by twenty in width, having sides ten or twelve feet in height, with a row of stout timbers, each some twenty feet tall, running through the center to support the ridge. In front of the house a bonfire was brightly burning, back of which sat a man by himself, who at our approach made signs for us to sit around him. As we stowed ourselves on the ground I asked MacCoy, one of the boat-steerers, who this fellow might be.

"His name is Kanker," he answered, "and he is the King's son, and can talk first-rate English when he is not full of awva, which he seems to be tonight." This I found true; and on becoming acquainted with him, as I shall presently relate, I found him to possess the most wonderful abilities of memory that I have ever seen displayed by any man.

The fire in front of us lit up the whole enclosure, showing on our right a high wall that joined the wall we had followed in coming here at right angles, just clear of the open gateway. On the left, coming in, it was about the same height, and ran back into the darkness. Just clear of the end of the house, on our left, the bushes cut off all view beyond twenty or thirty feet, and the wall was lost to view in the same jungle that ran to it from out of the darkness. Where we were showed like a large courtyard, with the house and front walls for background on our right, and on our left shrubbery.

Shortly after we had got into our places, coconut shells cut in half, answering as cups, were handed round by natives crawling on their knees to reach us. Looking into the one handed to me, I saw it was two-thirds full of nasty-looking liquid, almost as thick as mustang liniment. This I knew to be awva [kava]. I had seen awva before but never cared to taste it. This that I had in my hand may have been some of the finest that ever was chewed; and if to use the rule they who drink it say is the test, "The worse awva looks, the better it is when drank," then what I had in my cup must have been nectar borrowed from the gods.

I quietly set the mess partly behind me. I was not longing for an emetic just then, so I kept my eyes away from it, or no doubt I should have spilled myself all over the banquet ground. As it was, looking up to MacCoy, who faced me, just as he removed the shell that contained the dose he had been drinking away from his mouth, I saw strings of the vile decoction showing in the bright light, running down his moustache. I left the table pretty suddenly and went into the house.

Looking around me, after getting inside, I found the two ends were not enclosed. This, with a wide space left open on each side, one of which I had entered through, gave plenty of ventilation. A short distance apart, between the posts through the center of the house that supported the roof, were sticks stuck in the ground, on which were rows of the meats from a nut. These were nearly round and about the size of a large shellbark. These contained so much oil that, by lighting the one at the top first, it would burn freely and, when consumed, would ignite the next in row, and so on until the last one had burnt out. Each nut would give about the power of three candles. When burning they gave out lots of smoke, but it would not be noticed in a place so open as this was.

Along one side of the house, halfway between it and the posts and midway of the house, was a smooth piece of timber of some hard wood, having an oval shape on the outside. Two-thirds around it the inside had been hollowed out, leaving a shell from two to three inches its full length, which was some twenty-five feet. The rounded side lay upwards; and when struck with a piece of ironwood used for that purpose, it gave out a note that sounded more like the croak of a bullfrog with a cold than any other musical sound.

Some eight or ten men and women were in one end of the house, preparing the provisions in liquid form. This was done by breaking off, from a cluster of spider-shaped roots, a piece the size that could be conveniently placed into the mouth and chewed. After thorough mastication, it was removed from the mouth in a wad that looked like a bunch of dirty strings, only a little more so. This was thrown by each one into a large open calabash, and when it was full a man took it to one side and added water to the mess, manipulating it for a time. When he had succeeded in bringing the water to as sickly looking a porridge as possible, he would withdraw his hands, streaming with that mustang-liniment-looking affair. Next he took a piece of the fibrous substance that is found attached to the top of the coconut tree, around the trunk close to where the stems of the leaves join. This looks something like a hair sieve, and will strain anything about as well. Placing this over a part of the calabash, he strained this most delectable nectar into shells held out to him by man or woman.

Oh, ye gods! To what lengths will not humans resort for the gratification of vile passions? This stuff has to be drunk as soon as made, I have been told. Some of the machines manufacturing this compound, as they were quite good-looking girls, might help the stuff go down better; and that was all I saw about the process that was not too disgusting.

I went out again into the square. A number of natives could be seen on their knees with heads bowed down, around the entrance to the enclosure. On my asking what that meant, the boys told me they could come no farther until Kanker had made a sign or told them to move, as it was death for any native to move backwards or forwards after coming in his sight until he gave them leave so to do; and then as long as he was in their sight they could move in no other way but stooping as low down to the ground as they could make progress. This custom I found to prevail, and a more abject class of natives I never saw. Everything belonged to the rulers, nothing was exempt.

On a signal being given by Kanker, the natives at the gateway came crouching close along the wall as far from where we sat as possible and crawled into the house. There must have been forty of them at least. When all had got in, Kanker, half-stupefied from the awva he had imbibed, staggered towards the opening in the house, saying to us, "Come on, boys."

We followed him in and found, on getting inside, a row of natives seated along the hollow log, each one having a short stick in hand. and on the opposite side, facing them, another row of men and women with garlands of leaves around their necks; this being about all the dress they did have. Well, the night was warm anyhow. All bowed low and kept their heads down when Kanker came in and took a seat that had been prepared by placing on the ground a number of fern leaves. I did not know but some of the poor devils would unjoint their necks before the heartless Kanker gave the word for the band to strike up.

The first burst of the music and song was heartrending. I commenced to feel for the top of my head to see if it had followed the hair. After a time I recovered from the shock, and I then could understand what I had heard about persons becoming used to horrible sounds and thinking nothing about them.

Becoming somewhat calm, I took a look at the actors and found that their arms and bodies were moving in sympathy to the song they were howling with voices like mad bulls, though not so loud sometimes but what the log music could be caught. The fellows pounding it were doing their level best, with their skins shining with perspiration like a porpoise's hide when he springs out into the air.

If it had not been for the infernal din raised by them, the sight would have been more pleasing. The song that created the most disturbance had words over and over again that sounded like:

Ah f-a-r a-w-a-y—f-a-r—a-w-a-y,
A-h—f-a-r—f-a-r—a-w-a-y.

At times they would stand erect, all except the drummers, their bodies leaning forward, backwards, sideways, stooping halfway down, raising up, arms to right or left, above the head or pointing down, in perfect keeping with each other, as perfect as machinery.

I suppose a side issue was given for our benefit when six men and six women stood up opposite each other. The band struck up, and catching step with its strains they skipped towards each other with hands raised above the heads, palms outwards. Meeting halfway, the palms of the hands were brought forward together, making but one sound; then down on their naked thighs on the outside, and back again, something like the boys and girls do when playing "Pretty Polly Hopkins."

After repeating this a number of times, accompanied with the usual screeching that must pass, I suppose, for singing, back they would step into their places. Then taking some other queer freak into their heads, they would kick out first one leg nearly at right angles to the body, drop it, take one step forward, and send out the other. Coming together in the center at the right moment for the woman's right leg to be raised when the man's left came up, each stood with the other's leg in his crotch for a few moments, then made a stern board to places again. At times they would appear almost as if no bones could exist in those parts. As most of the time during this exhibition the women faced us backwards, a good chance was afforded for us to see the full development of their muscles. In some of their movements they were almost as supple as the Sandwich Island women, who are noted as being the leaders of the world in that art, or pastime.

This and other kinds of entertainment were kept up for some hours. There was quite a display of different things in the way of eatables, such as breadfruit, yams, a poor kind of taro, fish, wild pigeon and bananas. These all had been cooked in the ground, but no one took any of it except ourselves and the chiefs. During the time the programme was being carried out, awva had been passed around freely. Some of our boys and the 4th Mate were as drunk on it as Kanker was, who lay snoring in a drunken slumber, leaning against two or three of his slaves.

One of the dishes served that night I ate more heartily of than anything I ever have had compounded for me by such natives, before or since. The boys called it pudding. Well, it certainly did look like one, and a rich one at that. The arrangement was placed before us resting on a number of bright green leaves. These leaves, some 18 inches long and carrying their size well towards the rounded point at the end, had been placed around the pudding when bringing it to serve. The ends that fastened on top having been cast off, these were laid down to make a clean, inviting resting place for it. It was about the size of a large ball in a tenpin alley, snow white, without the faintest sign of dirt on or around it, and had the rich look of ice cream. I ate so hearty of it that I was ashamed of myself.

Kanker told me afterwards that such are only made for him or his friends, and no one is allowed to mix them up with his bare hands. The pudding is made of breadfruit, a certain sweet kind of yam, bananas, and the meat of the young coconuts when they are just soft. It is handled with a spoon, all moistened with the juice from sugar cane pounded up and squeezed through some of the coconut fiber; then made into a ball, covered with leaves of a certain kind, and baked in the ground for hours. When taken out and all the leaves from the outside removed with much care it is wrapped up as we saw it placed before us.

After finishing my share of the feast, I filled my pipe. Lighting it, I took a look around me. Some of the natives had disappeared. What were left had gathered in one end and were singing something that sounded much better, as one could hardly hear it. Most of the women had also gone, and so had all the boys that the awva had not sent to sleep. The 3d Mate had disappeared about the same time one of the Prima Donnas had, but I do not say that that had anything to do with it. The last nuts were burning out on the sticks in the ground as I took a final look at the wreckage in the house and started to go on board.

I lit my lamp, which I had put out and stowed away when first coming there, and took my way towards the boat by the path I had come. As I passed by those wonderful walls, my mind was made up that my first Sunday should be in part devoted to having a look at them.

Next morning the Mate roused out what men there were on board, and by sunrise the rest made their appearance. Some of the boat-steerers and crew looked as though they had passed through a bad winter—and poorly housed, at that. By noon we had stowed down the water and started another raft of empty casks on shore; and then I was put to a job that suited me the best of anything while we lay in that place. The Mate wanted pigeons shot, to give us in the cabin fresh meat of some kind for the table. As wild pigeons abounded in the woods, he inquired which of the boat-steerers could handle a gun. None of the others seemed to care for tramping through the woods with a double-barreled shotgun, to shoot pigeons or anything else, and, as I was most eager for such sport, it was voted that on me rested the duty of supplying the demand.

I took with me plenty of powder and shot and a native boy about twelve years old to carry the game. In the morning, after breakfast, a boat would be sent to land me on some point where solid ground existed on the large island, clear of the mangrove trees; and it would come for me in the afternoon on my coming in sight and discharging my gun.

Well, I used to have barrels of fun wandering among

the immense, strange tropical growth of trees and shrubbery. I found plenty of wild pigeons, but took fine care not to shoot so many in any one day that it would interfere with my shooting the next.

One of the days when out for pigeons, I came out of the woods onto a beach, clear of trees back of it, except for a grove of coconut trees under which lay a number of nuts that had ripened and fallen from them. As it was about noon, I chose a comfortable seat in the shade. Sitting on a slight mound made where the roots of one entered the ground, and resting my back against the trunk, I had a fine cool seat. I took out my lunch of salt beef and hardtack and began to eat, feeling as happy as a lord.

Nothing disturbed the quietness of the surroundings (I had not brought the native boy with me) except the attempt of a bird, now and then, amongst the trees farther back, to sing a low note. Perhaps frightened at his own voice, he would stop almost as soon as he started. Even the surf on the beach, as the sea was calm, would roll a comber in, start to curve and break, but tumble itself into a heap and send a small wave rolling a short distance up the white sand beach that could be heard no great distance away. These put me in mind of a mother raising a threatening hand to her child, and, after raising it quickly to the full extent over her head, slowly lowering it and saying as it reached her side, "I—am—a—good—mind—to-o-o."

I had been gnawing away on my hardtack and salt junk for a bit, when I saw one of the coconuts some little distance away from me roll over. I stopped eating, for I was startled. It turned completely over; but not seeing any more movements in it, I commenced to eat again. Shortly after I caught a chill, and it went all over my whole body. Even my hair seemed bristly. I certainly could hear plainly the rustlings of the dry coconut stems and leaves that thickly covered the ground around me, and could see some of them move.

There was not wind enough to move a feather, let alone one of those stems that would weigh four or five pounds. I lost my appetite right away. I do not think I finished chewing what was in my mouth, but dropped it into the lower hold as quickly as possible. I began to believe that the supernatural did exist when the climax took place; for I saw a dry coconut not more than thirty or forty feet from me fly into the air a foot or so (it had been lying on a bunch of leaves and stems on a slight ridge) and come tumbling towards me. I made up my mind that if the Devil was not here, it was about as good a place as he could find.

Grabbing my gun, pigeons and accouterments, I started to leave things just as I had found them. I would have humbly made apology for intruding to any and all things, but not knowing how it might be received, I did not attempt it. By going out on the beach and following it around, I could get within hail of the ship, or at least where my gun could be heard, by ten or 11 P.M. To be sure, every step I took, my foot would sink in the soft sand to the ankles, but I should not run any danger of "malaria," as I might by traveling through the woods for an hour and a half the other way.

I think this was about the first time I had thought so much about my health and the danger from malaria. I also about made up my mind to tell the Mate, when I got on board, that the malaria had taken such a hold of me that I did not think it would be safe for my health to shoot any more pigeons, unless two or three of us went together.

I had not made but a step or two from where I had left the balance of the lunch I had no use for, when I saw before me the most wonderful crab I ever have seen before or since. In his two immense claws he had a coconut, and he seemed trying to pull the husk off it. This explained the mystery of the moving coconuts. All thoughts of malaria were banished from my mind and in my eagerness to watch his movements I made too much noise, which caused him to drop his dinner. He was suddenly disappearing under the stems and leaves. Raising my gun to my shoulder, I pulled the trigger; but on pulling the wreck of him from under the stems, where he had partly succeeded in concealing himself, I found the shot had almost destroyed his shape. As near as I could decide from the pieces, he must have been about eight inches long, clear of two heavy short claws, one sharp-pointed, the other more stublike. These denoted that he might be capable of taking the husk off an old coconut, breaking the shell, and eating the meat. I was told afterwards that this is their principal food, and that these crabs have been known to ascend a tree and cut the stem of growing coconuts when none could be found on the ground. They live mostly on dry land, only going into water at certain seasons of the year; are very shy and only pugnacious when cornered; and they have been known to crush a dog's legs in their claws. The natives will not eat them, as they say they will dig up a person who has been buried and eat the meat from the bones.

Another day, I brought down an immense animal, bird or vampire. These things I had seen, now and then, through the openings in the trees, flying back and forth from one high tree top to another. The shot broke one of its fins, or wings; and when it struck the ground "thud" with only one of those things attached to it, the young imp of a dark-complexioned gentleman, standing just below, started off into the woods away from it, yelling at the top of his voice, "Dibalo!" He meant, as I found out afterwards, that I had shot the Devil, and until I had left the thing some distance behind on the ground, he would not come near me again.

On my approach to the thing, which was lying partly on its back, it struggled to right itself up and make for me, showing a shining row of sharp white teeth in a mouth stretched to its limits, with two long fangs projecting from each row of teeth, like a dog's. The head of this arrangement was in shape like a fox's, the eyes small, black, and running over with a devilish look that made me a little careful not to get within the reach of his teeth. His ears were sticking up straight from his head, which was three inches long, I should say, and covered with a short reddish hair like its body, which was in shape something like a squirrel's, and perhaps ten inches in length. Where the forelegs should project, long slim whalebone-like bones extended, having three joints ending in numerous fibers. These all ran downwards except two at the first and second joints. They were larger in size and ran upwards, having a clawlike hook projecting half an inch or more on the ends. These answered to hold the beast in place when he hung on the limb of a tree with all sail furled, looking like a bundle of black stockings on a line to dry. A short outrigger of the same kind ran out each side, where the hind legs would be. Over this network of bones and fibrous matter hung a loose, greasy, black, nasty-looking membrane. This formed the sail, when spread.

To put it out of longing for bananas or pawpaws any more, I gave it the charge of shot remaining in the other barrel of my gun. I found out afterwards that my fears were groundless about its biting me, as I was told they can be handled without fear. The natives, however, are superstitious about them, and will not go near one if they can help it.

Among the things the natives do believe in, is the existence of the Devil; and as far as I could sound the depth of water they drew in that direction, that was all they did believe, anyway, in the dim and distant. When one of the head men (the common man has no place here or in the future, with them) has a dream that his Satanic Majesty figures in, he assembles everybody at one end of this small island, on which the King and leading dukes live. (There are no duchesses or ladies here; they have other uses for the female women in this hell hole of depravity.) With loud shouts they form a close line stretching from water to water. Drumming on calabashes, pounding with a stick in one hand on a joint or two of bamboo held in the other, blowing huge conch shells through small holes made in the spiral ends to fit the mouth, yelling and screeching, they slowly advance towards the opposite end of the island until they reach the beach. Then, jamming together so that a cat could not well get through the line, they advance into the water until waist deep. After splashing the water furiously for fifteen or twenty minutes, they have succeeded in removing the Devil from their midst—as the dream was supposed to tell he had been.

If this line of march is broken by a man's falling down, or a gap being made in any way, then all have to return and form over again, as through such an opening the Devil has found a way to dodge. If a dream happens on the very night after these poor devils have, the day before, put him to flight, the same religious worship is repeated. Need missionaries? Oh, no; why should such saintlike devotions be interfered with?

During the time our ship lay there, they had one of these sublime and soul-stirring damnable times; and the dreadful din they made lasted in our ears for a week afterwards. We thanked our stars that no mistake had been made in the grand rally, as, if another had followed the first within a day or two, it might have rattled the royal trucks off the masthead.

One day, when out cutting ironwood poles, we came to a small village, and the sight of the people in it was perfectly terrible. They were simply being eaten up alive with the most loathsome of diseases. The state some of them were in was so sickening that I hurried away into the woods and cursed the white man who had turned loose this horrible thing among these poor helpless people. The sight of those in that village, where they had been put out by themselves to be slowly eaten up by the disease, haunted me for years.

WHEN I was on shore one Sunday, and he perfectly sober, I made it a point to meet Kanker. He stood about five feet, ten inches in height; a perfect form. His eyes, although a little dim from drinking awva, sparkled with spirit and intelligence when talking. His profile was fine, his age I should think to be about thirty-two or three.

I questioned him, asking how he became so well informed of events, persons and matters that were in all parts of the world where he had never visited. "If you had missionaries here," I said to him, "with schools, then I could understand something about how you acquired some of the knowledge you possess. But even then, I can hardly see how from that source you could know so many things that you do."

He informed me that he would tell me all about what I asked him, by and by; but he, right then, wanted me to answer a question about his allowing the missionaries to settle there. Some had wanted to come, but he and his father had refused to allow them to land, as they had been told by a number of captains and the crews of ships that had come here, that the missionaries would, if he allowed them to land, take all of any value from them and leave them poor. And the men who had left ships here, and had lived with them, also had told them no tobacco would be allowed to land, and that the women would not be able to get Jew's-harps, fishhooks and many other things.

"Now, I like you tell me what I do," said Kanker. "I see you no all same plenty other man. You no care stop all night shore, you no drink awva. You not tell me lie, I thought so," was the finale of the question he wanted me to answer.

For a few minutes I did not reply. He also was silent, but looking at me steadily out of his beautiful, dark-brown, intelligent eyes. Sailor as I was, with many of the faults that accompany that life, I could not answer him in any other way but what I thought was the truth, and best for him and his island home. I think that for a few moments the thoughts in my mind were as deep as ever I have had, before or since. It seemed that on me hung perhaps the lives of many poor girls, and the ruined health of numbers of young men who would come there in ships. These and many other thoughts flashed with lightning rapidity through my mind; but in a short time I roused up, and with a laugh to myself, if any of the boys saw the reckless, fun-loving and devil-may-care Nelt in the guise of a preacher, how they would roar with laughter, I said to him:

"Kanker, I will tell you what I think is best for you and your people. Let the missionaries come, give them a piece of land, and help them put up a house. Do all you can to get the natives to attend the schools. You can attend the school yourself, and they will be only too glad to teach you. You are too bright a man to kill yourself with drink. Those white men who run away from ships here are bad men. Have nothing to do with them. They are the ones that want nothing good.

"Now, I will tell you what the average men will do who come out here as missionaries, and give you some reasons why you should do all you can to help them. But I warn you to look out at the same time that they do not take advantage of your good nature, and serve you as some natives on other islands in the Pacific Ocean have been—more so perhaps in the Sandwich Islands, where missionaries have been allowed by the natives great privileges, and rule those islands with more power than the King.

"There must have been on this island, not many years ago, large numbers of people. What has killed a great many, no doubt, is the disease brought here by the men on ships visiting this place, with no remedy known to you to cure it.

They slowly die with it, as numbers are dying today, over on the big island where we saw them the other day when we went to get wood.

"The missionaries will bring medicines to cure such diseases, and will do all they can to stop any more being afflicted by putting a taboo on what is so common here now. They will teach the women and girls how to make hats and other things instead, that will give them such things as they need. Instead of all your people dying off, in a few years there will be more inhabitants on this island than there are today. One thing more I will say. That is, you, who want to know so much how to read and write, and also about the things of the world, should do all you can to have them come, so as to reap the advantages that you will have by their information."

He had been watching me closely during the time I had been rattling off the answer to his question; and he held his gaze fixed on me for half a minute after I finished. Then, jumping up, he said, "You no lie! I no hear sailor man speak all same you do. Me tell mekenery come quick!"

He no doubt kept his word, as hardly a year elapsed before a Mr. and Mrs. Snow settled on the island; and from all accounts they were among the better class of that kind of people. I have found, in my dealings with the missionary element, that while all perhaps may be good, you can still find some a great deal better than others.

Here was the most remarkable native I have ever met; and I must speak of what Kanker told me in regard to his acquiring so much knowledge of the outside world and persons. This prince (for prince he was, even if his royal robes only consisted of a red woolen shirt that had come from some whale ship, and a narrow piece of cloth around his loins, knotted in front, with the two ends passed between his legs and tucked through the standing part where it passed across the small of his back) had the most remarkable memory of any man I ever heard of. I have read to him slowly quite a long sentence, and had him, after I had finished it, repeat, almost identically, word for word of what had been read; and asking him some days after about it, he could repeat it over again. For instance, he could repeat from the life of Washington, Bonaparte, Wellington, passage after passage; and about the battles they fought he could tell in detail, as far as I know, just as they had been written.

He shamed me one day by asking how many presidents (and what were their names) the United States had had since it became free from England, and then repeating the number and names, as far as I know, correctly. Many accidents and large fires he told about that had taken place in different parts of the world. He had books and papers in numbers piled up in one corner of his house, and begged all he could from us, as he did from every ship visiting his place.

His manner of getting information from them, so he said, was to have someone read aloud to him from them. One man, he told me, was landed from a ship, sick. He took care of him and had him in his house. The man lived some three or four months with him before he died, leaving a large number of books, amongst which were histories and lives of prominent men. "This man," he said, "read to me many things, but the best of all the reading I liked was the battles"; and his eyes would flash like a hawk's as he said, "I would like to do that." Some persons had added to newspaper articles the most astounding things, and he would sometimes repeat these, causing uproarious laughter from those that heard him, much to his astonishment. Some of his words were so far out of their course in the way he pronounced them that a six-fold, patent-purchase set of blocks could not hoist them into place.

After supper I took another turn up to Kanker's house,

taking with me a few books that I could spare, and a few trinkets. He was sober and seemed pleased to see me. I read some out of the books I had brought him, and he seemed to hang on every word I read. It seemed a pity that he could not be taken to some place and taught what he so much wanted to know, and not have his fine intellect destroyed by the effects of intoxication and his mind debased from the associates that at times he would have around him.

I asked him if there was any information he could give me about what these walls were built for? What was their purpose? He replied that he did not know when they were built, or what they were built for; and the only thing he could tell me was that many, many years ago, their tradition said, men of large stature lived here and were of lighter color then he or this people were. Before I left him to go on board that evening, he said that anything I wanted that he could get for me, he would do. I told him that I was built like the rest, but that I never drank liquor or did anything else for pleasure when the chance was of its hurting me; and I felt convinced, from what I had seen and heard since we anchored here, that when we went to sea a good deal of medicine would be used before we had a well crew—or words to that effect.

Calling a native (that I named Jim, afterwards) inside the doorway, he told him something in native, at the same time pointing to me, and the fellow backed out after he had finished.

On my rising to leave, I bid him good night. Then he told me that he had ordered one of his best men to attend me whenever I came on shore, and to obey no one else, not even himself, when I was in sight. He was to see me on board and be on the watch for me on landing, day or night; and whatever I might ask for, I was to get the best to be had. I thanked him for his kindness, hardly knowing what to think of it. On leaving the house, sure enough, the fellow kept close to my heels until I got into the boat to go on board, and every time when I went on shore after that he would pop up alongside me shortly after, if he was not at the boat to meet me. He was very faithful, and on the eve of the ship's departure I made the poor fellow so happy in presents I gave him that he fairly danced with delight.

One Sunday I thought, in my walks on shore, that I would measure a tree which I had seen growing amongst the rocks one of the high walls was built of. Taking a few fathoms of spunyarn with me before leaving the ship, I was prepared when I arrived at the spot to see how much it would girth. I gave one end of the string to Jim, who scrambled over the stones that had tumbled out around it, from the side nearest which it grew, and brought it around the trunk to me. Hauling it taut and marking the place where it showed its distance around on the spunyarn, I measured it when I arrived on board, and found it showed sixty-five feet; the height and spread of its branches was enormous.

This tree had started its growth long years after the wall had been built. There could be no doubt of that, as it grew in a portion of a wall that had been tumbled down enough to allow the shoot it grew from to take root amongst the stones which had formerly been part of the wall that in places was now twenty to twenty-five feet in height. Say the tree was five hundred years old, then certainly the wall must have been built before that; for no seed could work itself from the unbroken part through the stones and reach the ground. Even if it had, the want of light would not have allowed the seed so dropped to germinate. This being the case, the stones had to scatter enough for light and heat to allow a seed to have the benefit after reaching the ground; so, when I say the wall must have been built one thousand years ago, I do not believe I am out of my Lat. or Longitude.

What could have been the purpose of the people who built these things? I wondered about this time and time again, sometimes standing on top of a wall fifteen to twenty feet in height and almost as much in width, and built of stones so large that they would, in many instances, require heavy purchases to raise them into position. How those who built them succeeded in placing stones weighing a ton or two in the places they occupy, without some such appliances as we use today, would puzzle, I think, anyone who saw them, as it did me.

SOMETIMES I went with stores for our Captain's use from the ship to Captain Hussey's house, and often met him. Captain Hussey was a tall fine-looking gentleman, and one would hardly have taken such a pleasant soft-spoken person to be a man who had cut himself off from the world and all ties of civilization, except now and then a wandering whaler; but such was the case. He was afraid to seek his home where his loving wife looked with weeping eyes towards the ocean when his ship returned without him, in which three long years before he had sailed away, after kissing her tears away with loving words of hope, on his return, to bring back with him enough so that he no more would have to leave her and the little ones.

But now how different! He, on one of the most blighted inhabited islands in the Pacific, amongst a people so low that a good dog would, no doubt, be better as a companion; and she, thousands of miles away across land and sea in her little home on Nantucket, with many tokens of his love around her to remind her how many happy hours they have passed together in it, since he, a young bright-eyed sailor, on the eve of sailing as captain in his first ship, brought her, a rosy-cheeked young bride, to this home.

The cause of Captain Hussey's choosing to isolate himself in this Godforsaken spot was peculiar. His cruising ground, when on board ship, was mostly on the Line. He had been fairly successful up to the time his ship had been two years from home. Calling at one of the islands in the group that was friendly, he found a good trade for coconut oil could be made with the people on the island, if he would leave on shore casks to be filled with oil, and come for them in a few months.

He landed the casks and sailed away, cruising for whales until the time was up for his return to take the oil that the natives would have ready for him. As he was making the island, a canoe came on board with the Chief in it, who told him a large lot of natives had landed from another island a short time before, and had killed quite a number of the men who were guarding the oil and driven away the others. They would soon take the oil away, as they held possession of it and also that part of the island where the oil was; and he could get it in no other way except by sending some of his men with guns to help drive off the intruders.

The casks that he had landed were worth some $100.00 or so, and the trade advanced about as much more, which he did not like to have stolen from him. So, after consulting with his officers, it was agreed to assist the Chief to recover the oil, the men agreeing to the same. A force of men were landed with Captain Hussey at their head, armed with muskets, pistols, and a small cannon that was on board the ship. Joining the natives, who had been assembled by the Chief on the opposite side of the island from where two or three hundred raiders held the village where the oil had been gathered, they marched across in the night and took the other party by surprise, just at daylight, by pouring a volley into the huts where they were sleeping.

The battle was soon over. The raiders had had no thought of firearms being used, as nothing but clubs, spears and swords made of shark's teeth were the implements of warfare used by them on either side. Rushing with frightened yells from the houses towards the beach where the canoes lay, as many as could shoved off and paddled for their lives in the direction of the island they came from, some five or six miles distant. A number were killed with clubs and spears by the natives of Captain Hussey's party, before they could get into their canoes and escape.

The recapture of the oil, with some canoes and a large lot of implements of war, was the result. Paying the Chief for the oil collected, Captain Hussey took it on board.

In visiting Sydney, some months after, Captain Hussey heard that an English man-of-war was in search of his ship to take him prisoner, as the English government claimed jurisdiction over the group that this island was situated among.

This affair, and one other he was concerned in, which I will relate, caused him such fear of what might happen to him if he did not conceal himself that he settled up the business of the ship to the perfect satisfaction of the owners, when she arrived home in charge of the mate, by getting to some out-of-the-way spot but little visited.

To go on with the statement of the other affair: a mutiny occurred on his ship, led by a man who had been at the bottom of many troubles that took place before this outbreak, which was brought about by the men refusing to lower the boats for whales unless the demand they made of having the ship furnish a better quality of food was complied with.

The captain reasoned with them all he could, and told them, amongst other things, that the food they complained of was all the kind there was on board the ship, and that when the ship went into port where he could get better, he would do so. Trying in vain to have them turn to their duty, he ordered them to take the places they belonged in and lower the boats; and they refusing, he went into his cabin. Coming out with a rifle in his hand, and loading it in their sight, as they were standing forward on the forecastle deck and could see every movement he made, he said to them, when he had finished ramming home the ball in the gun, that they had five minutes to think the matter over and do as they had been ordered. If at the end of that time they had not returned to duty, he would shoot down the first man he told to take his place if he did not obey.

Laying his watch face up on the binnacle, he took a turn back and forth on the quarter-deck. Looking at the watch when the five minutes had passed, he picked it up and put it into his pocket, only saying to them, "The time is up." Taking the rifle, he advanced to the forward part of the quarter-deck. Raising the gun to his shoulder, he leveled it at the breast of the ringleader, who had been standing with his shirt bosom stretched open with both hands, daring him to shoot.

Captain Hussey took a deliberate aim and told him by name to take his place. Instead of that, he jumped to one side as the gun went off; and the ball struck and instantly killed one of the best young men of the crew, who had been led by this wretch to open act of mutiny. Captain Hussey had not time to change his aim, as the other man, when jumping aside, quickly caught and swung the other into the place he had occupied.

A few seconds after this man was shot, every one of the mutineers sprang to the places they had been ordered before this tragedy took place. Among the first was the cowardly villain who had caused the trouble. Every man was now ready to lower boats or anything else he might be told to do.

Seeing the mutiny was ended, the Captain ordered everything made fast, and the poor fellow who lay dead on deck to be laid out for burial, by sewing him up in canvas with weight enough to his feet for sinking him. He was launched overside into the sea, so closing the second act that caused one man's death and a fond husband's exile.

Both these cases were settled by the courts in the States, on evidence furnished by the logbook and the crew of the ship: the first case on the rights of volunteers helping to protect their own property. The shooting of the man in the act of mutiny was held justified to save discipline and the welfare of ship and owners. The exertions of Mrs. Hussey to bring the thing to an issue were untiring, but by the time it took to reel off the log the amount of red tape required, she had expended about all her means and the Captain was lying like the man he shot, at the bottom of the sea. These facts, concerning the cases and his death, we heard at about the time we started for home, two years after we met him at Strong's Island.

The manner in which he met his death I will state as we heard it:

Some six months after our leaving Strong's Island, a brig on a trading voyage called in there. Her captain had been drinking so hard that he was rendered almost helpless, and the amount of trade she had on board was trifling. After Captain Hussey had talked with the captain of the brig about fitting her up and trying to catch a few whales, he consented. As there were two whaleboats on shore, belonging to him, Captain Hussey put them on board, with some line, and other things he had got from a whale ship or two, that would answer to take a whale. He got together a crew of natives, and with the few men that came in the brig, he started out on a cruise amongst the group.

The information of what occurred after he left the island for the cruise came from some of the natives who sailed on the brig with him, as none of the few white men who were in the vessel when he left ever were seen again. It seems they had taken a whale or two and had also collected some coconut oil up to the time when, at an island, the name unknown, trading for oil, the crew, aided by the natives from shore, attacked and killed Captain Hussey and every white man on board. They then ran the vessel ashore, plundered her of what they wanted, and set her on fire.

A few months after Captain Hussey sailed on his last cruise, letters came from the lawyers who had conducted his case through the courts, stating that he had been cleared of all charges and could now return home. There were also brought, by the whale ship which conveyed the letters, some from his wife, expressing the fond hope of their soon meeting. Sad was her heart, when looking day by day for his appearance, to be met with tidings that no more would they meet in this world again. *

* Starbuck's *American Whale Fishery* records that the ship *Planter,* of 340 tons, sailed from Nantucket in 1847; and that her former master, Capt. Isaac B. Hussey, shipped on board the brig *Wm. Penn* and was killed in a mutiny, November 6, 1852.—*Ed.*

Out of the Frying Pan

AFTER spending two weeks at Strong's Island, during which time we had taken on board all the wood and water that we had available space for, and also given the ship a coat of paint on the outside, we took up our anchor just at daylight one morning. With a gentle breeze blowing fair from the large island, we stood out through the reef to the sea, and before the wind failed, had an offing of a few miles. Taking the sea breeze when it sprung up about eight A.M., we trimmed our yards to it and stood towards the South.

About two weeks after we had left that hell's kitchen, we one day sighted a school of whales. A better chance for four boats to get each a whale could not well have happened; but we were able to lower but two boats, on account of so many of the men being laid up with sickness contracted at that headquarters of disease. The pain and sight of those who suffered was sickening, making the few untainted happy to think the will to abstain had saved them from such trouble, which in some instances would last to the grave.

Those that were well had to attend the sick. The ship for

a time was more like a hospital than anything else. The cooper and myself were the only ones in the steerage that were capable of helping our roommates there, now and then lending a hand, as well, to help the Mate and 2d Mate attend cases amongst the men forward. We only got two whales out of the school we lowered for, and it took us longer to cut them in than it would have taken to cut in four at other times. Those whales made about twenty barrels each.

Cruising to the South for a month or six weeks without seeing the spout of a sperm whale, with part of the men who had been on the sick list able to do duty but looking very pale and exhausted, found us in the Lat. of 15° 40′, Long. 180° 00′—E. or W. of Greenwich, just as you might take it. We were on the Meridian directly North of the easternmost point of the northern island of the Fiji Group. Our position was about one day's sail from it. Owing to our having been so long on this cruise without being able to get but little in the vegetable line (not much of such could be had in any places we had touched since leaving Eoa about six months ago, Strong's Island being only good for wood and water, and exhausting the medicine chest of certain kinds of medicines) had caused scurvy to appear among the men. We wanted to take a cruise in New Zealand before going into the Bay of Islands, where plenty could be had; but unless we did get something to stop the increase of scurvy, we should be compelled to put in there and shorten our cruise.

The Captain made up his mind to touch at one end of this island, inhabited by the blackest kind of Negroes and the most cussed set of cannibals in the Pacific. Telling the Mate that one end was safe to trade on, but the other end would massacre anyone that got within reach, but, as we must have, if possible, something, we would try to get it, he said that he had never tried to land there, and would not attempt it now if it could be avoided without too much loss of time in seeking some other place that had what we needed.

The next day, after he had shaped the ship's course to make the land, about half an hour before sundown, "Land ho!" was cried from the masthead. "Right ahead!" And high peaks of a mountain could be seen, when the sun went down, towering above the horizon eight or ten thousand feet, some thirty or forty miles away. Taking in the light sails and hauling up the fore and mainsail, we stood in towards the island all night. At daylight the north point of the island was abaft the weather beam, about six or eight miles off. Making all sail, we stood along on the port tack for half an hour longer. The Captain coming on deck gave order to tack ship. When we had the ship on the other tack, and the yards sharp up against the backstays and shrouds, her head looked up, so as to bring the point of land two points on our weather bow. Standing along for an hour, we passed it within a mile or so; and as we drew out by the land quite a village could be seen in a deep bay on the main part of the island to the left of where the point, or peninsula, joined it.

On seeing this settlement the Captain gave orders to the Mate: "Have some whale's teeth brought on deck, also some blue cotton cloth, some lead cut into small pieces, powder made up in little paper packages, and some fishhooks." These were to be put into the waist boat when she had her loose craft taken out and was ready to take him on shore. As I steered the waist boat, it devolved upon me to go on shore when she went. I did not much fancy taking a chance of being eaten up by those niggers, but got my boat ready when the order came to do so.

The ship stood off the land until the village bore four points on our weather quarter. The wheel was put down and she came in stays. Filling away on the other tack, we soon ran in near enough to lower the boat, in which the Captain went. Before we lowered he told the Mate that he would land if he found we were on the right end of the island, but if not, he would soon come on board.

Shoving off from the ship, we pulled for shore, the Captain having the steering oar. After pulling a mile or more we came to a reef that extended across the bay where the village stood, back of a white sand beach at its head. This reef joined on the shore to our right and extended as far as the eye could reach along the land to our left, with no chance to land unless we could find an opening into the bay of smooth deep water that we could see in front of the beach on which the village stood.

Pulling along the reef a safe distance from the breakers that were tumbling over each other in wild confusion on its outer edge, we came to an opening just wide enough to allow our oars to pull through without their blades touching either side. The reef on our right, as we came through the passage, soon ended, leaving a smooth bay of deep-looking water, large enough to hold three or four ships with a chance to swing at their anchors; but the reef on our left continued almost in a straight line from the passage until it joined the beach a little to the left of the village of perhaps fifty or sixty houses, some of which could be seen back in a valley that was between the two ranges of hills.

The distance from the inside end of the passage to the beach was perhaps one-half or three-quarters of a mile. About two-thirds of this had been passed over when the Captain spoke up quite sharp and said, "Tie your oars!" We obeyed the order, raising the blades from the water by resting one arm hard enough on the run to bear it down over the gunwale inboard, and keeping it steady with the other hand.

Looking over my shoulder, I saw two natives with a couple of bunches of bananas which they had set down on the beach a short distance from the water. Looking back towards the Captain in the stern of the boat, I could see a puzzled look on his face. The next minute he said, "Peak the oars! I do not like the look of things here."

Each man shoved the handle of his oar into cleats made for that purpose, which gave him liberty to look around. The rule, after that order is given, is for the boat-steerer to get on his feet and stand on the platform in the bow of the boat, called head sheets, and be prepared to carry out any order that can be done in that position.

As I stood looking towards the shore, two or three more natives came out from the houses, waving green boughs, and one holding a small squealing pig. I thought this looked all right, and wondered why we lay here when we had come inshore to get some of the same things that were now offered us.

I soon found out that the Old Man was too old a bird to be caught with chaff, for in a few minutes he said to me, with a half-sigh, "Nelt, we are on the wrong end of the island. Those cusses mean murder, but we need something for the sick men and I think by using precaution we can get something and get out of here with our lives. You see, there is not a woman in sight, and hardly any men. No doubt in my mind that the minute our boat touched the beach, fifty or one hundred of the black devils would come tearing down and brain every one of us before we could shove the boat off. However, you get the boat warp ready and we will pull in a little nearer to the beach, where the water is about deep enough so that it will leave their head and shoulders out. This will give us the advantage and if they bring off anything to trade that way, we will buy it. If not, our boat is afloat and under control of the oars and we will get out of this."

Pulling with four oars to within a hundred yards of the beach, we found that the depth of water was as required. When he had the boat in the depth of water wished for, we were alongside the reef that joined the shore. It was nearly dry; and its side was as square up and down, almost, as a wall. By throwing the bight of the warp over a spur on the reef that stuck up a foot or so, the boat could be held until ready to go, when, by letting go the running end and pulling on the standing part, she would be clear without a man getting out to cast it off. Making the end fast, after hauling around the spur just enough so that a push now and then with the boathook kept her head clear of the reef and a man amidships on each side with paddles kept her from swinging broadside on the reef, we were in position to receive company.

Those few natives on the beach seemed to be watching our movements with much interest; and waiting a few minutes after we had completed our maneuver, they let out a yell that made the hills echo, which in a few moments brought out from the bush and houses such numbers of the black imps that it reminded me of the swarming out of ants from a nest disturbed. To show that the Captain was right, very few had stuff to sell.

Holding up a whale's tooth, the Captain made signs to bring off the bananas and other stuff. (Whale's teeth are to the Fiji natives the most valuable of anything you can offer them. Ten small teeth, at the time we were there, would purchase outright any woman or girl. Fifteen of the same kind would buy the most stalwart native. You could eat, kill or take them away. The coin value of that number of teeth, as they were sold on the market, would be about $1.50. A blackbirder would have been all right for a big load by having five hundred dollars' worth of whale's teeth.)

Taking on their shoulders a bunch of bananas, a big bundle of taro, yams, or whatever they had for trade, a number of them waded into the water and slowly made their way to the boat. When they got alongside, a tooth would be held up for a certain amount, or a fishhook or some cloth. With a nod, their trade was thrown into the boat and the thing offered was passed to them.

The natives who brought off stuff for sale, I noticed, did not return to shore for more, but hung around the boat. I also noticed quite a number of them working down the beach and out on the reef towards the boat. Pretty soon some of those had arrived abreast of the boat and they were all carrying spears and clubs. By the time eight or ten of them had got opposite the boat, they standing as close to the edge of the reef as possible, those in the water became more offensive than they had been before, trying to shove the boat against the reef and taking hold of the gunwales with their hands.

The Captain, seeing it was time to get out of this, commenced to get his trade out of the way, when two or three natives got hold the gunwale forward of amidships, and gave the boat such a lurch that it almost tumbled the Captain out of the boat into the water. He swore loudly and sung out to me, "D—n you, Nelt, keep the devils from upsetting the boat!" Having the boathook in my hand, I made a motion to strike those who had their hands on the gunwale. Two or three removed theirs but one of the number (who I think must have been a leader, if they rank according to looks) would not let go his hold. Turning upwards one of the most horrible, devil-looking faces, with a diabolical grin he showed almost every tooth in his head, the upper front ones filed sharp like the teeth of a saw, no doubt to give him a more ferocious appearance and a better chance of tearing human flesh. The cussed look he gave me, and what the Old Man had said, so aroused what little bulldog I had in me that, let follow what might, I would hit him one for some of the poor devils he had pulled the meat off the bones of with that set of sharklike teeth he seemed longing to insert in some part of my body.

Bringing the heavy handle of the boathook, with the most strength I possessed, across his two hands grasping the gunwale, as he tried again to tip the boat, I heard the crack of bones; and I think no mistake occurred that every finger that handle hit would be off duty for a time at least.

I almost wanted to laugh (but just then there was not time) to see him take those hands from the gunwale, open the top of his head, and try to stow them away in that fiendish mouth of his.

When I hit him he let a yell out, and at the same time shouted out some words. For a minute everything was quite still. As soon as I had hit Cupid, I reached for the standing part of the boat's warp, letting go the running end, intending to haul it clear of the spur; but by this time a rush was made by those on the reef towards it, some treading on it, and one pulling on the part that was attached to the boat. Seeing that in a minute more the boat would be pulled to the reef and we at their mercy, as quick as one could count "two" I pulled the knife from its sheath in the head of the boat, that is carried there for instant use; and hardly had they taken in the slack of what I had gathered before the boat was free, leaving them gaping with the rope in their hands.

When I cut the rope, I sung out, "Stern all! Stern like hell! Or those hellhounds will have us!"

The discipline and practice of a whaleboat's crew told a whole lot right then, as hardly had the order been given before the men dropped their oars and with the same motion sent her astern in time enough to keep clear of those in the water, who made a rush about the time the fellows on the reef captured the boat's warp. There would have been short work made of us in a few minutes longer. Our boat being turned over, we would have been at their mercy in the water.

It may sound a little funny, my speaking about what I said and what I did, with the Captain in the boat. But you have to take into consideration that the hitting of our saw-tooth friend, and cutting the warp, and the sterning-all took place while the Old Man was picking himself up from amongst the trade where he had been tumbled when the last roll was made on the boat. It did not take him many seconds to get on his feet and grasp the steering oar and quickly lay the boat around, pointing her head towards the passage in the reef through which we had entered this pretty-near graveyard for us.

The tide had fallen since we came through the opening. There was plenty of water for us to go out, but the reef on our right was quite bare as far out as the inside of the passage; and over this the blue-and-black-bellied devils were running and yelling, shaking their clubs and spears when they could spare the time in the race between us to see who would win the day by reaching it first.

If they succeeded in reaching there before we did with their spears and clubs, which they throw with accuracy, the chances were dead against us, for the passage being so narrow every one of us would be speared or clubbed before we got through it. The natives, stumbling over the reef, had a little advantage of us in distance, as the edge of the reef on which they were running lay nearly in a straight line, and we had to make a curve in our course that would take us out of reach of their spears being darted into the boat. However, we were making the boat fly through the water faster than they could get along over the uneven surface on the reef. Still, we cast anxious eyes now and then towards those howling devils who were so eager to make us into roasts.

The Captain said but little. Now and again in a low tone between his teeth he would say, "Pull hard, but do not break an oar, for that may cost our lives."

Well, I have done tall pulling, before and since that morning, but I think none of us ever beat the pulling we were doing then. The pinch came when we had to turn the boat's head for the passage, as in doing this we pulled quartering towards them. This could not be helped. They saw at once that it gave them the advantage and that it was now or never with them. Letting out fearful yells at every leap they made, they seemed to keep pace with the boat for a few minutes. Then we dropped them, all but one. He was a powerful cuss, some distance in the lead of all the others and with his spear poised for darting, and it seemed he would string some one of us up. Making a spring, he landed in a hole that tumbled him down, about thirty or forty feet from the stern of the boat, sending his spear flying some feet ahead of him. Quickly gathering himself out of the hole he had stumbled into, he grasped his spear and sent it towards the boat with all the force he could, but he fell short and the spear went into the water with a hiss, half a boat's length astern of us.

The boat's head entered the passage about the same time the fellow with the spear tumbled, and the game was ours. How it might have ended if that fellow had put his spear through the Captain, who had the steering oar, is hard to say, as a little disarrangement with the steering oar at that time might have cramped the oars on one side of the boat or the other and given the other imps time to get some work in with their spears. As it was, we went out through the passage with a rush of foam under the bows that made a sound like a young cataract. Some of them ran along the reef until the wash from the breakers on the outer edge stopped their progress, and in desperation they hurled spears and clubs that fell far short of us.

We pulled a safe distance off from the reef and rested with our oars apeak. Every man in the boat was winded. While we rested, we sent showers of curses after the black imps who could be plainly seen working their way to the shore, turning now and then and shaking a raised spear or club towards us.

We had now a chance to see how much stuff we had got in the way of trade. Two or three small pigs, some dozen or two chickens and about half a barrel of taro and yams was the size of it; hardly enough to pay for taking off to the ship, which was almost hull down, seaward.

The wind had hauled more to the North since we came on shore, so a course could be laid, with a sail standing hard full, along the land to the S.E. end of the island, some fifteen miles away. The Captain, after talking about our narrow escape during the time we were resting, said to me, "I think we will not take the little stuff we have in the boat on board, but set the sail and work up to the other end of the island, which no doubt is the end we can do some trading with, if any. So set the sail, and as soon as the ship can make us out we will keep off for the other end."

The ship, heading in with the wind two points abaft the beam, soon caught sight of us. Seeing they had the run of the boat, we kept off; and with all the oars pulling and the sail drawing, we shaped our course along the reef that ran from three-quarters to a mile from the shore, where the high mountains and hills ascended, in places, quite abruptly. Some few green valleys could be seen now and then, between ranges of hills that extended back, but these always ended against a bulkhead of sharp peaks. It made our mouth water, with a hot sun pouring down its red-hot rays on our unprotected heads, at the thought of rolling in the grass under the shade of some of the many trees that could be seen as we glided by.

Looking towards the quiet lovely spots of tropical scenery that could now and then be seen, reposing so peaceful, one might expect to see lambs and children sporting on the patches of bright green grass. But instead, we knew that around those peaceful spots lurked the most devilish cannibals unhung, watching to steal each other's children, drag them in the bush, and eat them, when the craving for human flesh was unsatisfied from other sources. As each village or bay is at war against the other, and all prisoners are eaten, they only make a raid on one another's children now and then.

We had covered about two-thirds the distance between the point of the island we started for and what might have been our graveyard, when, looking towards a man on the beach who seemed trying to attract our attention by waving something in his hand, the Captain told me: "Peak your oar and see if you can make out what that fellow is trying to do. Your eyes are younger than mine. See what you can make of him."

Peaking my oar and getting on my feet, I could see him much plainer than while pulling. There was no question but what he had a hat on his head and a shirt on his back, but if he was black or white I could not tell.

Turning to the Captain, I told him the result of my observations. "Tie your oars," he said, at the same time letting the sheet of the sail loose from the cleat.

When the boat stopped, the thing with a hat ran along on the beach for a short distance, then headed towards us. He could be seen picking his way over the reef, wading in the water halfway to his knees. Then he stopped and commenced to wave both hands. Just then, in his frantic endeavors to attract our attention, his shirt flew up enough for me to see white skin under it. I sang out, "That is a white man." The Captain asked me if I was sure. When I told him I was, he kept the boat off for where he stood, and although he seemed standing behind a row of breakers that no boat could go through, as we approached near him he waved with his hand to sheer to the right. In doing so we shot by a point of the reef that ran parallel to the main reef far enough to form a little bay. Luffing around it with a few quick strokes of the steering oar and holding water on the two-oar side, the boat's head came up to the edge of the reef in smooth water that would not have broken an egg.

The white man (that's what he called himself) came tumbling over the box of the boat with an old straw hat on his head and a shirt covering his body. This was the extent of his wardrobe. He was the most filthy thing of that name that could be found alive by hunting the world over, as to be any more filthy and breathe would have been impossible. His villainous-looking, mummylike face, when he grinned on passing by me to go aft where the Captain was, caused me almost to think the fellow whom I had, a few hours before, interfered with wearing kid gloves for a time, was the more respectable looking of the two.

Backing out with the oars until clear of the point, we got on our course again. The man told us that he lived on that part of the island we were now heading for; that the natives there were all right for trading, and we could get plenty of taro, yams, bananas, sweet potatoes, hogs, chickens, and ducks, and that the natives were friendly. But he startled us with the report of a whaling vessel, belonging to Sydney, making a mistake, some short time before, about the right end of the island to trade on, which landed two boats in the bay we had been in that morning, and every man had been killed and eaten.

He said that he saw our ship off that place and was afraid that another boat's crew or two might be cut off; and when he saw the boat, from the high land, coming along the reef, he felt better. He did not say whether it was on account of the trade he might get from the ship, or whether it was because we had not been eaten by human sharks.

By five P.M. we made a landing, and after trading for a lot of stuff we made arrangements to come on shore the next morning and get the balance of what we needed. By dark we got on board the ship, which had followed us up.

After getting supper, I told the other boat-steerers about our little picnic, but all the satisfaction I got out of it was that they would pity the poor unfortunate black devils that had to take me for their share. On my saying that this was nothing to joke about, they replied: "Of course, it would have been no joke to the poor hungry orphans that had eaten you in innocence and bliss."

"Go to 'H,' or Halifax," was my shot at them as I turned in to my berth.

The next morning the ship was close in with the land. After an early breakfast, two boats were lowered and a lot of trade was put in them, consisting of whale's teeth, fish-hooks, powder, lead and some cloth. We pulled in with the boats to the landing on the beach where some two or three hundred men, women and children were assembled. Amongst the number was our dirty-faced acquaintance of the day before. He seemed to have some effect on the black horde by keeping them away from the boats until we had them hauled up on the beach; and the trade was carried up to an open shedlike place where we were to do our trading.

The fun soon waxed fast and furious. Hogs were brought in squealing, slung on a pole, and tumbled down in front of the shed. Chickens, yams five or six feet long, taro, bananas, and many kinds of fruit were soon piled up in unlimited quantities. In two hours' time we had bought as much stuff as we could have use for.

The natives were a surly looking lot of the blackest kind of cusses, but were very peaceful, not showing any disposition to molest or steal anything. They were as a rule of large frame, and as no dress covered them, a fine view could be had of their muscular forms. Some of the females could have stood as black Venus, with perfect grace. None of the boys were tempted, though, by charms displayed, to lose any of the fishhooks they had brought on shore to trade for shells.

The Captain gave the fellow, who had been of great help to us, some cloth, an ax, two or three small whale's teeth and other things in payment for his kindness. This man was a convict who had escaped from Sydney by stowing away on a whale ship that had visited that place. He informed our Captain that he had been living on that part of the island some eight or nine years, and had two or three wives; and that the natives never troubled him, as he had been tabooed by one of their priests, when he first landed. Some of the boys said that the reason he was not eaten by them was on account of his being so dirty. He told some horrible stories about their feasts on human flesh.

Before night we had taken on board what recruits we had purchased and made all sail, heading to the South on our course to the New Zealand whaling ground. In a few days after leaving the Fijis, the men who had the scurvy commenced to recover, and in a week or two were entirely well. So, by the time we arrived on our cruising ground around Sunday Island, our crew were all well except a few who had not entirely got over the blight that had struck them while at Strong's Island.

OUR BAD luck seemed to follow us, for after cruising on that ground for nearly a month, without seeing a whale, one day we were hove to in a heavy gale of wind, showing nothing but a close-reefed main-topsail, fore-topmast staysail, and main spencer, when the ship drifted into the largest body of whales that it ever has been my fortune to see. There were whales to the windward, whales to leeward, whales ahead, whales astern, whales everywhere. The most wonderful sight of all was to see them tumbling about in the water. When the big waves would break, the one on its crest would be rolled partly over on his side, and with a quick turn of his flukes, right himself, then shoot down the side of the hill of water like a boy sliding down the side of a snow bank on a sled, and seemingly enjoying the sport as much. Some of the whales would come very near the ship, not seeming to show any fear of us. Perhaps instinct told them that in a gale like this no boat would venture an attack on them. Every man was on deck watching the antics of this vast body of whales with longing eyes. The excitement of seeing thousands of dollars' worth of oil all around us, and we powerless to take any, rendered the crew from Captain to cook frantic.

The officers implored the Captain to let them lower the boats, but with set teeth and face a shade more pale than usual, he steadily refused all entreaty for an hour or more. At last some five or six whales, in a body, came close to the ship on the lee beam. "Look at those whales laying there, asking us to come and catch them," yelled the 3d Mate, dancing up and down in his excitement. "Oh, for God's sake, Captain, let me lower my boat and try to get one!"

The Old Man took a look at the whales lying five or six ship's lengths dead off the lee beam, and then turned his face to the windward, gazing with a troubled look at the heavy mass of dark, lead-colored clouds that hung low over the raging sea. For ten or fifteen minutes he steadily kept his eyes in that direction. All hands knew how intense must be his feelings, as he more than anyone in the ship had the most at stake, both in standing and interest of all whales taken. During the time he was looking away from the whales that lay to leeward, hardly was a word spoken by officers or men. All had respect for his deep anxiety. Not a man but what knew that to lower a boat in a gale like this would be to take a great chance of its never being able to live in such a sea, or ever get back to the ship.

He turned around, holding by one hand on the lanyard of the main swifter, and called to the 1st, 2d and 3d Mates, asking them if they still wanted to lower for the whales? They promptly replied that they certainly did. He sung out to the men belonging to the three boats, "If anyone does not wish to go in the boats, he need not do so." But as one voice they all said they were willing to take the chance, and wanted to follow the officers.

He shook his head, turned to the windward, pointed his finger at the rolling seas, and said: "It is no time to lower a boat, with seas running like that, and if we did not need oil so badly, I never would consent to allow a boat to lower from this ship. But you all want to try and get a whale, and perhaps you would blame me if I do not allow you. But if lives are lost, never will I forgive myself for consenting to your wishes."

He then told his three officers, "Under no conditions strike but one whale. Bear in mind that if we can get one in such a gale as now is blowing, we would be the most fortunate men, so let every boat help the one that strikes first. Lower away! And be careful!"

As we were hove to on the starboard tack, the three boats on the port side were to leeward. It would have been impossible to lower any boat to windward without smashing it to pieces against the ship's side.

By careful management the boats were lowered and we got clear of the ship, but it was the most tumbling about of whaleboats that I ever saw before or since. We could see now much better how easily the big seas could pick a ship up and tumble her about; and once or twice it seemed as if she might be thrown on top of us like an avalanche of hull, masts and yards, before we got out the oars and away from her side.

Getting the boat before the wind and sea, we went flying with the oars in the peak cleats. We needed no sail, as one the size of a postage stamp would have needed a reef in it if it had been set. The whales that we had seen so close to us in a body had gone down, but plenty of spouts could be seen to the leeward, and as it would have been useless to try for any whale that would have brought the boat broadside on the sea, we kept running before it. The boats were but little distance apart. Still, it was only now and then that one could be seen, and then only when on top of a sea. We ran by quite a number of whales but did not see them in time enough to sheer the boat.

The Mate happened to get one just right and put two irons into him. He was ahead of us and the whale he struck never attempted to run or sound, but lay on top of the water with the sea rolling him around almost like a log. He was but a short distance ahead of us when he struck, so we came rushing down and alongside of him. A quick heave on the steering oar, by the 2d Mate, shot our boat alongside him under his lee, and this gave me a chance to send both irons into him; and one hit his life and set him spouting blood. When my irons hit he took a little sound but soon came up again with other whales around him.

The 3d Mate came up but mistook the whale we were fast to, and his boat-steerer darted his irons into a loose whale and did not find out the mistake until the whale he had struck started to run off before the wind, away from the Mate's whale, who now was spouting thick blood. The Mate sung out to Mr. Griffin that he had better cut our line, as his whale was nearly dead, and follow the 3d Mate, who now had found out his mistake and ought to cut his line; but, as he did not seem willing to do so, he might need help.

We cut our line, bent on two spare irons that are always carried in the boats for like occasions, and started in the direction we had last seen his boat, but the surroundings were such that he had disappeared in a few minutes after he had struck the whale he was fast to. Soon we were by ourselves, with nothing in sight but the ship, which could be seen at times when we both mounted on a high wave together. When down in the hollow of the sea with walls of breaking water around us, the dark storm clouds over us hung so low that the tops of the mountain waves would appear, by a little stretch of imagination, to pierce them.

The sea would now and then almost cause the boat to broach to, as we had it on our quarter. The direction we now were steering brought both wind and sea there. For some time we had been following the slick on the water made by the oil and blood flowing from the wounds made by the irons in the whale the other boat had struck, but in a short time we lost trace of that. We struggled along for an hour or more, sometimes with two men bailing the boat to free her from the water that came into her from breaking seas.

About this time it struck me that we are doing as fine a d—n fool thing as men could well do and not try hard, either. Still, I did not like to speak my thoughts, for an officer is supposed to do all the thinking and the men to do the acting. I had been standing up in the head sheets (as the 2d Mate had ordered me to keep a lookout for any trace of the missing boat) for fifteen or twenty minutes, when he said to me, "Where do you suppose that other boat can be?"

My answer was, "It would be hard to say. It may be capsized or the whale may have turned a little off the course we have been following and we have passed one side or the other of him. He might be within a short distance of us and we would be unable to see him unless both of us happened to be on top a sea at the same time."

"Well, I hardly know what to do," was his response.

"If I headed this boat I know what would be done and that as quick as possible," I ripped out pretty bluntly.

"What do you mean by that kind of talk? And what would you do if you headed this boat?" was his question, in rather a cross manner.

"Head for the ship and do the best to reach her that I possibly could," was my reply to his last question.

"You would, would you, and leave the other boat and men to perish, maybe?" was jerked out spitefully.

"Look here, Mr. Griffin, you do not know where that boat is, no more than I do! She may be anywhere, and if we keep on going in this direction until we get so far away from the ship that dark would overtake us before we could get on board, what show would there be for us? And then again, if that boat is swamped and the men are right here in the water, could we take them into this boat now? Without so overloading it that we would all go to Kingdom Come in no time? You know, as well as I do, that it is only by the utmost care on your part that this boat keeps right side up, as she is now. Let us hope they are all right and will do the best for their own safety, as we should do for ours. I have had my say and I am willing, as I know the men are, to follow any course you wish to pursue."

For a few minutes he made no reply. Then he said, "I am sorry we cannot find them and I think you are right in what you say, so we will try to make the ship."

Laying the boat around head to the wind and sea, we took our oars and commenced pulling. The ship was some six or eight miles to the windward. We had made out to see her take the Mate's whale alongside and luff again to the wind. As she did not know what had become of us or the 3d Mate's boat, all she could do was to drift in the gale, hoping that we might reach her. Men were aloft on the yards and in the riggin', doing their best to catch sight of the lost boats. The Captain was in the topmast crosstrees, sweeping the troubled waters with his glasses, swearing both loud and deep at his weakness in allowing the boats to lower, and his orders not being obeyed. Flags were set in various parts of the ship and whipping themselves into rags in efforts to call the wanderers home.

We strained every muscle to keep the boat under headway enough to meet the seas that came rolling down on us like small mountains and sometimes would almost conquer the headway of the boat. If that should happen we all knew our fate, as the sea would turn our boat over like an eggshell. We had succeeded in making our way through the water towards the ship, but the ship in her drift had shortened the distance more by far than anything we had done; and after some two hours of pulling so, she was only a mile off, and all were beginning to breathe more easily, when I heard an ominous roar ahead of the boat. Looking towards the 2d Mate, I saw his eyes were set on what I knew to be, by the sound, an "unfortunate sea," so called by sailors. (These seas come at times from no one knows where or how, and sweep a ship's decks, smash bulwarks, and do any amount of damage, and for days perhaps may not be seen again.) I did not look over my shoulder to see what was coming, but did as the 2d Mate ordered us all to do: "Pull for your lives!"

We bent the oars in the effort to give the boat headway enough to pass the hell of roaring waters that were rolling down on us like the sound of some vast cataract. It struck us, the foaming water around us like a vast snow avalanche. The boat had raised her head and met the first crash, then trembled like a thing of life for a second or two, trying to recover and hold her way. Finding too much against her, she dropped her head away two or three points, and shot astern with such force that the blade of the steering oar caught in the water, thereby raising the loom so quickly that it caught the 2d Mate and tumbled him over the side into the water, but not so far away but what he caught the after oar and got back into the boat.

We had swung almost broadside to the sea and our boat was up to the thwarts with the water that had broken into her. A small sea striking us now would seal our doom. The steering oar was useless for the time, and if she was not brought head to the sea instantly, she never could be. Before the 2d Mate had crawled back in, I had got the boat pulled head to the sea and knocked the heads out of two water kegs and was bailing the water out of the boat with two men, and pulling three oars when the next sea came for us. As good luck would have it, we rode it without taking in much water.

The 2d Mate soon got his steering oar all right, and when we had a chance to bail out the water we improved it, and soon were fighting our way to the windward again. The ship did not see us until we were within a half-mile of her, and then only when we rose on a sea. Everybody was much pleased to see us, and you can believe me when I say we were pleased ourselves. The first question asked of us was, "Where is the other boat?"

As we had not seen the other boat since the Mate had, we could not give any information. The Captain was wild over the thought of the loss of her and the boat's crew, for he as well as the rest of us made sure she was swamped.

About an hour after we arrived on board, a man on the foreyard sung out, "Here's the other boat! Close to, on the weather bow, coming down before the wind."

Sure enough, there she was, not more than a dozen ship's lengths off, tearing down before the wind and sea to cross our bows, safe and sound, but looking as if any minute the sea must swallow her surely. She shot across the bows under the jibbooms so that it took but a slight sweep of the steering oar to send her under the ship's lee into smooth water. Laying her short around, the 3d Mate brought her under her tackles, and watching for the proper moment, both tackles were hooked on and willing hands soon ran her to the davit heads, with such rattling of the patent sheaves in the blocks as is seldom heard.

Getting in on deck, the 3d officer was called by the Captain to give an account of his act in disobeying the orders that had been given before the boats were lowered.

"I have only this to offer. In striking the whale I mistook

a loose whale for the fast one, and when I found out the mistake, I thought that I might save that one too. For not cutting my line, I disobeyed your orders, and that was wrong, I know. It shall not happen again," said the 3d Officer.

"I hope it will not," said the Old Man. "What became of the whale you struck?"

His answer was, "He lays dead, to the windward of us. Soon after I struck him he hauled on the wind and I got under his lee. As he hardly moved I hauled the boat close on to him by the line and set my lance over his shoulder blade. He only spouted two or three times, then he rolled over on his side dead. The whale made us a good lee, the oil from his wounds made the water smooth, and no seas broke around us until we cut clear from him to reach the ship."

"You must have passed the ship no great distance ahead," said the Old Man.

"No, I did not. We could see the men at the masthead and expected you would see us. We had the waifs set, one on each end of the boat," was the reply.

The Captain asked him if he saw our boat. He said he only saw us once, and we then were to the leeward of him, on his lee quarter.

"Well, it is too bad we could not have saved the whale you struck, but it would be hard to find him now, if we could carry sail to work the ship to the windward. But as we cannot, we must put up with it and feel good to think you are all safe aboard and that we have one whale, even if it is a small one. So go below and put on dry clothes," said the Captain.

As it was very near dark, the order was given for supper, after which all hands turned in but a boat's-crew watch. Nothing could be done, as the gale was fierce as ever, except to watch the line that held the whale veered out on the ship's weather beam.

Such times as these, with a whale alongside, the only safe way to hold him is by veering him out with a line some thirty or forty fathoms to the windward of the ship. This is done when the wind is blowing so hard that it is impossible to cut a whale; and to keep him alongside fast by the chain fluke would be likely to part the strongest chain or cut the ship's bows down from the hole where it comes through to make fast at the windlass bitts. A line of two-and-one-half-inch rope is securely fastened to the whale and the gangway taken out, so the rope can be as low on the line of the water as possible, to prevent chafe. (Of course, it must be understood that all whales are taken to the starboard side of the ship.) The end of the rope is taken around the main bitts, or some firm stanchion, with a round turn and a man tending it to slack it at times and to haul it in at others (or, as sailors express it, veer and haul). Sometimes the ship will be hove by the sea faster than the whale. At such times the rope, being taut, would be liable to part if it were not slacked properly; but as a rule the ship and whale drift very fairly together.

NEXT morning, all hands were called at daylight. Cold, dull-looking clouds were tumbling over each other to the windward in the same mad fashion as the day before, and the wind was blowing equally as hard. Owing to the oil and blood from the wounds in the whale spreading over the water, the sea to the windward did not break; but outside of the slick the whale made, the sea was breaking with a roar of fury.

The Captain told the officers, "We can do no cutting on the whale today." So after breakfast all but a few of the men went below.

All through the night following, the gale continued. The morning following, the gale not abating, men were set to hauling in on the line that held the whale, to bring him near enough to the ship so that some cuts could be made in his blubber to let out the gas that was in his body. This had swelled him to a third more than his natural size, and if not let out it might burst him open and bother us about stripping the blubber off when the chance offered. The first cut or two, with the long-handled spades, sent the pent-up gases, blood, and stench into the air quite a distance. The wind catching it and blowing it inboard gave those in line the full benefit, and soon it penetrated every part of the ship. After we had cut some scarfs in him, on the same lines that the blubber would have to be cut when stripping it off with the tackles, he was veered out again. Some of us got old spades and cut sharks that came in reach, for by this time there were any quantity around the whale; and many a beggar was sent rolling over and over into the depths of the ocean.

On the third morning the gale had moderated a little, so the order was given to haul the whale alongside and make the attempt to cut. We put the fluke chain on him, hauled his head under the after stage, and wore the ship around on the other tack, so as to bring him to the leeward; for it would be impossible to cut to the windward with seas running like they now were.

The tackles were got over the side and the first piece was raised. The officers who were cutting had to abandon the stages that had been slung over the side and stand on the rail, as the whale at times, when the ship rolled heavy to leeward, would be above the ship's bulwarks.

After we had been cutting a short time, it seemed, when the ship fetched a heavy roll to the windward, that her mainmast would be jerked out of the decks all standing, as the upper blocks were fast to the head of it by large hemp straps. After making a fearful lurch to the leeward, the ship suddenly rolled to the windward, and bringing the heavy falls taut, she gave such a quick jerk to the mast that her whole frame trembled; and if the piece that the lower blocks were fast to had not torn off, I believe the whole concern would have come down—head of the mainmast, main topmast, and all above it.

The Captain sung out, "There, that will do! Send down the tackles. We must try cutting him across decks, or not at all."

To cut across decks was hard work on the waist gang. We boat-steerers knew too well what that meant. (Still, there is no strain on the ship.) The two upper blocks of the cutting tackles are lashed on the opposite side of the deck, directly abreast of the gangway, one running-part of the fall of one block leading forward to the windlass, the other fall leading aft to the capstan on the quarter-deck through heavy snatch blocks, to make the falls lead straight on windlass and capstan. The piece of blubber is raised over the plankshare in the gangway and pointed down the main hatchway, and, with a couple of men down there to cut off pieces as fast as it comes down, this makes it seem an easy way to cut in a whale.

The officers with their spades cut the blubber on each side as the ship rolls to the windward and turns the whale. As the piece bent over the plankshare brings a strain on it when she rolls to leeward again, the piece, now being slack, will slip in with a light strain on the falls. But as it would not do to allow the lower blocks that hold the piece to cross the opposite side of the combings of the main hatch, on account of tearing the blocks to pieces, each block can do only about ten feet of blubber. Then they have to be lifted and overhauled by hand; and hauling those big falls through is about the most backaching work I ever had on board of any ship.

We occupied the whole day in cutting in the body of that whale. The head we had to leave until the next day, and by that time the weather was good enough to send the tackles aloft and hoist it in, as nothing but the body of a whale can be cut in across decks.

The whale, when tried out, made about thirty-five barrels. The share of each green hand out of that would be about five dollars; a rather slow way to get rich. Still, when we take into consideration the Glory attached to a whaleman's life, one perhaps ought to be happy. One should not want the whole world.

We spoke several ships but none had had such poor luck as us. This did not cheer anyone much, on our ship. Strange, how things will happen about catching whales. Two ships cruising the same ground, and every few days being in sight of each other, one will succeed in making a good season's catch and the other will hardly get oil enough to grease his pots.

Covered with Glory

DURING the next month we cruised about Curtis Rocks, and Goat and Sunday Islands, but were not fortunate enough to see any whales. We landed on Goat (or Macaulay's) Island and caught a few wild goats. We also stopped at Sunday Island, and found a white man settled in the only place of anchorage around it that would be safe for a ship; the bay being open to S.W. winds. The man who was settled there had two wives, but was very hospitable to any captain who might stay overnight. Sometimes two captains would spend a night on shore with him, playing checkers, or some other quiet game.

One day we had been lying becalmed since daylight. Turning out from my afternoon watch below at four P.M., I started aloft to relieve the boat-steerer who had the masthead at the main from two until four. When he passed me on his way down, I standing in the topmast crosstrees, he told me nothing was in sight and he did not expect there would be, ever again. In a joke I told him I was going to raise a whale before suppertime. I had no thought of doing so but wanted to say something. He went on down the topmast riggin' and I crawled up the topgallant rattlins into the topgallant crosstrees. Shortly afterwards the 2d Mate relieved the 3d Mate, who had been at the masthead from two to four, as the boat-steerer had whom I had relieved. We were both half-asleep and the glassy look of the ocean did not tend to help us brighten up much.

I suppose we had not been to the masthead more than half an hour when I saw a spout that perhaps was about two miles off, but as it looked like a finback spout and was no good to us, I said nothing. In a few minutes I saw it again, and for want of something better to do I reached my hand around back of the royal mast for the spyglass. Seeing me reach for the glass, he asked me what I saw. "Nothing but an infernal finback," I replied, "but I think I will take a look at him." Slowly setting the glass on the right focus, I took a look in the direction I had seen the spout, and saw at the first glance something that caused me to drop the glass, for there lay a sperm whale plain in view. Nearly his whole length of back was above water, and it showed him to be at least a hundred-barrel whale.

"Great God!" I said to the 2d Mate, "There lays a sperm whale as big as a mountain. Sing out!" In my excitement, I forgot he had as yet seen nothing. He was in a minute as excited as I was, clutching hold of me and saying, "Where, where is he?"

"Never mind, he is there! You hail the deck!" was my wild reply. He yelled out as loud as he could bawl, "Sperm whale, sperm whale!"

The Captain, who was on deck, thought the 2d Mate had gone crazy and spoke out crossly, wanting to know what the matter was with him, hailing the deck in that manner; and where was a sperm whale? "Three points on the starboard bow, about two miles off," I told the 2d Mate, and he answered as I had told him. On receiving the reply, the Captain sung out, "Come down from mastheads!"

Dropping down the topgallant riggin' low enough to

catch the main-topgallant backstay with my hands and swing off, and then locking my legs around it, I slid down so much faster than the 2d Mate could travel down the rattlins that I was on deck before he had got more than halfway down the main-topmast riggin'.

During the time we occupied coming down from aloft, the Captain had given orders to get the lines in the boats, and hoist and swing. "Lower all four boats!" The boats dropped into the water with a splash, about together. Shoving clear, we shipped oars and every man lay back with a will at the word, "Pull Ahead!" and made the light cedar crafts fairly hum through the smooth water.

Now, no one but myself on board the ship had seen the whale. It was a queer thing, for it to happen that way; but the 2d Mate, in his excitement and flurry at the masthead, was not able to make him out before we left there, and no one could have been able to see him very well from the deck, even if they had looked, which they did not, before he went down. So the Captain kept pulling, and the other three boats followed, for some twenty minutes after I told the 2d Mate that we were pulling away past the place I saw the whale before he sounded. I was mad to think how much risk we ran of gallying the whale by pulling over him with our oars, and perhaps lose the chance of striking him by doing so.

Soon the Captain and the rest of us turned the boats around, for the ship had signals set to say the whale was between the boats and the ship; and back on our course we pulled almost as hard as ever. After pulling a little time, we made out to catch sight of his spout, but before we got very near to him he went down again, and did not seem frightened even if we had pulled over him, so I felt quite elated to think that the next time he came up one of the boats would be able to strike him.

As we pulled about to the spot where I thought the whale

went down, I expected the boats to stop pulling, but instead of that the Captain still kept on, and away followed the others. After standing the strain as long as I could, hoping every minute to see the boats stop pulling until the whale should come up, I could stand it no longer and spoke out in church by asking the 2d Mate if he did not think what we were doing was not d—n fool work? At that question he got red in the face and wanted to know what I meant by such talk as that to him.

My reply was, "I meant what I said! Just fool work we are doing; and it will end in our losing the chance to strike the whale, as you must know that where the whale went down is now some ways astern of us. For this time when he was up, every officer and the Captain also could plainly see him, and why we are pulling over him and away past, I cannot see. If going over him with five oars the second time, as we most certainly have, does not gally him, he is not to be gallied in such a way."

He seemed to think perhaps I was not altogether wrong, for his manner and speech were altogether different than at my first outbreak, as he replied that he thought himself that we ought to stop pulling, but that the Captain was pulling as hard as ever, and he must follow the other boats and him.

My answer to that was, "This sport might be kept up with some chance of winning a whale in the end if the time of day were different, but we have not time to allow that whale another sound. If we miss the chance to strike, this next rising, before he comes again the sun will be so far below the horizon that we would not be able to see him so far as we could hear his spout. So it is this time or never, and you head this boat and have a right to use your own judgment, and were I in your place I would not pull one foot more in the direction we are pulling, but heave up until he did show himself. If he comes up where the Captain seems to think he may by the way he and the rest are going, why, there is three boats' chance to strike him and we would not be missed; and if he should happen to come around here, then we could have a chance at him if we do not pull any farther away. But if we pull much more, that ends this chapter."

In a few minutes he said, "Tie your oars, I will take our chance here, hit or miss."

I never peaked my oar with more pleasure in my life than I did at that moment. Standing up in the head sheets, I watched the other boats pulling away for dear life until they could hardly be seen. But I was still so firmly of the opinion that I was right in my judgment that I whispered to the bow oarsman, who pulled the next oar to me, to keep a bright lookout astern, for I believed the whale would show himself there or nowhere. I had no thought that he would come up anywhere else. This whale had been about forty-five minutes in his former sound, and such large ones as he are very regular when undisturbed; so we began to look with anxious eyes as the time nearly expired.

The other boats had now pulled so far away that only now and then could one be seen as she rose on the ocean swell, when the after oarsman, a bright-eyed Cuban, set us all wild by almost jumping out of the boat and shouting at the top of his voice and pointing with his finger a little on the starboard quarter, "Hell and Jew's-harps, there he blows! There he blows!"

Sure enough, he could be seen, a short half-mile off, with nearly the whole length of his back above the surface of the water and his hump sticking up as big as a small-sized barn door.

The 2d Mate hove the boat around with the assistance of the oars about as quick as ever a whaleboat could be turned in that way, and headed for him, saying, "Spring, boys, on your oars. Spring hard, I tell you!"

After pulling a few minutes, I thought it would be right

for me to suggest another proposition, but I did not dare to until, looking over my shoulder at the whale, I could see he was lying headed away from us, and that the boat was rapidly approaching him. Thinking I might be taking too much to myself, I spoke to the 2d Mate, and touched him lightly on taking the paddles. He took my remark kindly: "Do you think we can surely reach him before he goes down?" And when I replied, "Yes and more," he ordered, "Oars apeak, and take the paddles."

It was a glorious sight, sitting on the gunwale of the boat, facing towards the whale that lay so still with his hump sticking some two or three feet above his back, which was hardly buried except when he slowly dropped his immense head and sent a swash of the sea rolling across it, after blowing from his spout hole his breath in such a lazy and quiet manner that the sound of it could not be heard more than a hundred yards away, calm as it was.

The boat, from the vigorous strokes the excited men gave the paddles, slipped rapidly through the water. Not a word was spoken above a whisper and every man used the utmost care that the blades did not strike the side of the boat. The boat made so little noise that I think a person thirty feet away with his back towards her would not have known when she passed him.

When within twenty or thirty feet of the whale, the 2d Mate in a low tone of voice told me: "Stand up." Laying my paddle carefully into the boat's bottom out of the way of its catching in the line, I took the first iron from the rack (called the crotch) and got it into position for darting. The next minute the boat's head was over his immense flukes, which could be plainly seen a few feet under the boat's bottom, beneath where I was standing. With the speed we were going, the boat shot like a dart by his hump. I raised the iron, and with all the force I was capable of, I sent it to the socket into the vast mass of blubber and meat that was now only a foot or two clear of the boat. The next instant I had planted the second iron within a foot or two of the first, just as the sun touched her lower limb in the western horizon.

"Stern all!" said the 2d Mate, as the men, seeing my irons go, had jumped to their oars after laying in the paddles.

We sterned the boat out of the foam the whale made when he raised his flukes fifteen or twenty feet above the water and brought them down on the surface with force enough to half fill the boat. He sent the foaming soapsuds in every direction on feeling the cruel irons buried two or three feet in his body. Rapidly throwing out the fifteen or twenty fathoms of stray line that is coiled away in the box of the boat, so no trouble can arise in sterning clear of a whale after striking, I made my way aft to take the steering oar, and the officer went forward to lance the whale.

On feeling the irons in his side, he struck one blow with his flukes and dove beneath the water; and by the time I had taken my place in the stern the line was flying with such rapidity around the loggerhead that smoke was arising from it quite freely, although two men were throwing water as fast as possible into the tub where it was coiled. The steering oar being peaked (this is done by raising the blade out of water and shoving the handle of the oar into a band of leather nailed by its ends, one outside and the other inside the gunwale, leaving space enough above for the handle to slip in), I caught two nippers. These are made square, about six or eight inches on the sides, of folded canvas sewn strongly together, to grasp the line when the whale is sounding, as the friction is so great that the bare hands would be burnt to the bone in short order, and the line run out in half the time it would otherwise; for after a little line has run out, the bight of it, if not restrained, would take out as much as the whale—more so, when the whale is making a running sound.

It is not to be supposed that by holding the line you can stop the whale. A ship's hawser of the largest size would not do that, for a hundred-barrel whale would no doubt snap that like a woman would a piece of cotton thread. With the nippers one in each hand, a double turn on the logger-head brought the boat's nose down within six inches of the water. The sea being so smooth, this could be done with some safety, but not so when it was anyways rough.

"You look out what you are about!" yelled the 2d Mate to me, raising up from clearing his lance from the side of the boat (where it is carried in cleats and beckets, with the head covered by a wooden sheath). "Do not box the boat down any more—you may turn her over." "Aye, aye, sir!" was my reply. Still, I held her down all the same, watching the men at the same time to see that perfect trim was kept, as at such times a sudden tip either way to the boat would be sure to upset her as quick as a flash.

About half the line had run out when I felt the strain on it ease up a little. "He's coming up," I shouted to the 2d Mate. "All right, let him come," was his reply. Shortly after, the strain was so much less I sung out to the boat's crew to haul line. They turned around on the thwarts face forward and clasped the line, and soon we commenced to draw in quite lively. "Haul away lively now, boys, he's coming up fast! We soon will have a chance to set his chimney on the blaze," said the officer.

In a few minutes he broke water with his head, sending the water in a cataract from it, and an affrighted spout that might have been heard a mile or more away. Snapping his immense jaws once or twice together, he dropped his head and out came his tail; and if he did not cover the sea with foam for acres around him (by striking it with his flukes, the sound of which was distinctly heard on the ship two miles away) about as quick as I ever saw it done, I am mistaken. The next move he made was to roll over and over, and so quickly that no person, without seeing it done, would believe an immense monster like that could move so lively.

"Now is our time," the 2d Mate yelled at the top of his voice. As we had gathered most of the line back into the boat, his words meant for me to place the boat near enough to the whale for him to use his lance. The 2d Mate was a bully fellow in such times. Some men would have hesitated, but he, not much. There was the whale and dark approaching. Although plenty of danger existed, there was a show to place a lance into his life that would win the prize to us.

"Take the oars and pull for him," I said, dropping the steering oar into place. The men sprung to the oars like heroes and soon had her near the leviathan, but he suddenly turned and swept his flukes out of water just clear of the boat's head. "Close shave," was all the remark the 2d Mate said.

Sterning two oars and pulling quickly ahead shot her alongside him, with the boat's head within six feet of his fin and over the most vital part of his body as he rolled back up to raise his head to spout.

"Bully boys," said the officer, as he raised his lance and pointed it down over the shoulder blade. And the next instant, shoving the keen steel head into the body and holding on by the pole, with four or five feet of the shank following the head out of sight, he raised it up and down as a person would a churn dash, then sang out, "Stern all!"

Drawing out the lance as we sterned clear of him, the 2d Mate turned towards me, saying, "We have him now," I had hardly told him that I thought so too, when the whale broke water (as he had settled on getting the lance thrusts from us) and sent out thick blood the full bigness of his spout hole. "See! See!" said one of the men. "His chimney is afire!"

We had been so busy that we had not paid much attention to the other boats, which now came pulling up with a rush, the Mate's boat ahead. He pulled straight to the whale, which lay like a log on the water, with his life blood pouring in thick clots from his spout hole. When near enough, his boat-steerer sent two irons into him that he hardly noticed, as he was near his end.

When the Captain hove up, a short distance from us, he said we had done well; that he would go on board the ship and send the boat back in charge of the 4th Mate, and for us to take the whale in tow, when dead, and that we must set our lights. (Each boat has a long slim keg, a little larger at one end than the other, which is lashed up on the underside of the covering boards that cover the stern sheets from gunwale to gunwale. In this keg are matches, candles, tinder, flint, steel, and a lanthern.) He would set lights on the ship to show her position.

Laying his boat around for the ship, he spoke to the Mate, telling him that, as a light air seemed springing up, if it was enough to give steerage way on the ship by the time the whale was turned up, to come on board with the other boats and leave our boat with the whale. Before he had got to the ship, the whale rolled over on his side, dead. We soon after had the lines clear of the whale, the iron poles pulled out from the sockets on the irons, and the hole cut in one end of his flukes, through which a line was put and doubled, thirty or forty fathoms, for hauling the whale alongside the ship to put the fluke chain on him. We had no more than finished this before it was quite dark, but with a clear starlight sky. As by this time the wind had sprung up quite a fair breeze, the Mate and the other boats went on board, leaving us with the whale.

In the course of half an hour after the boats left us, the ship came creeping along towards us, showing her sails and spars like a phantom ship. But little noise could be heard from her, the sea being so smooth that she hardly rose or fell on its surface. Getting in reach of our double line, her headyards were laid square, the afteryards braced up on the starboard tack, wheel put hard aport, and as she came head to the wind, all the jibs were run down and the Captain told the 2d Mate to run our line. This taken on board, we waited until the chain was put around the flukes and our line unrove through the hole in them. This being coiled away in the boat's stern, we pulled around the ship to our place under the tackles and soon had her fast on the cranes, the bully boat for that day.

I had not more than seen her grips well fast and other things about her secure, before I heard a loud call for the 2d Mate and myself. Answering, I was told to follow the 2d Mate into the cabin, as the Captain wished to see us. "Holy Moses," thinks I, "the Old Man would not be so mean as to get wild over the matter of our not pulling on with him! We have the whale, and we proved that he never would have been taken today in any other way."

As for me, I could not see how I could come in for any censure. First place, no one but those in the boat knew of the conversation that had taken place, and they had not had time to repeat it; and the 2d Mate would not for the sake of his own standing have it known that a poor boat-steerer had influence enough to cause him to disobey orders. I was willing to take any blame in regard to the matter on myself; but that would not do. So, in doubt what it could mean, I timidly crawled down the cabin stairs after the 2d Mate, who had gone before me.

At the foot of the steps I stopped, before I took any steps along the passageway that led from the forward to the after cabin, to hear a word or two that would give me the lay of the land, or, in other words, tell me what kind of humor the Old Man was in. Hearing the Old Man laugh sent my spirits up as high as the royal trucks in half a second, and with my hat in hand I boldly stepped into the cabin.

On seeing me, the Captain, calling me by the nickname

I went by, "Nelt," said he, "Come here. You have covered yourself with Glory today, and must take a drink with me." I thanked him and filled out of a bottle standing on the cabin table, a good tot of liquor. Raising it to my lips, at the same time saying many returns of the day, I drank it off, amid the laughter that seemed brought about by my remark. Making a bow to him and the officers, I went on deck feeling much better in my mind than on going below.

I went along to the waist where the other boat-steerers were looking over the rail at the whale lying alongside. One of them said, "Nelt, that is by far the biggest whale that has been alongside the ship this voyage. It was lucky that you spread the chances by heaving up where you did. What made the 2d Mate stop following the Captain and the rest of us for?"

My reply to him was, "Perhaps he thought we needed a rest, as we had outpulled you so far on the first stretch."

"Rest, nothing! I believe he must have had an idea we were pulling too far over him, and that's what made him heave up," was his next remark.

"If he did have that idea, it was correct, as is proved by what is alongside," was my answer. The other two boat-steerers both said that they thought all the time that their boats were a long way past the place where the whale went down.

Just then we were called to supper. As it was so late, extra plates were furnished for the boat-steerers all to eat at the same time the Captain and officers took theirs, so the conversation ended right there, and I never told my act of insubordination.

During the night the watch on deck got the cutting tackles aloft, and hoisted on deck the empty casks and blubber tubs, so as to clear the blubber room, for the whale was so large there would be too much blubber to have on deck in a case like this. The cutting spades were got down from their racks under the boats lashed bottom-up overhead, and everything was made ready to start cutting the whale at the first crack of day.

At daylight all hands were called and the work of cutting in the whale commenced. At noon the case, after bailing out of it fifteen barrels of pure spermaceti, was cut away from the tackle that held it up the side, end on, while the oil was being bailed with a bucket from it; and down into the depths of the ocean it sank, leaving nothing in sight outside of the ship of the huge leviathan, so full of life not many hours before, except a floating carcass of meat and bones some two or three miles off the ship's weather beam, around which were hundreds of sea birds, doing their level best to put out of sight what was above water, while sharks were doing the same to it beneath.

We finished cutting the whale in six hours of actual work. As he made one hundred and twenty barrels of oil, the time taken to cut such a whale could hardly be beaten by the best.

We extended our cruise towards the Three Kings. These are a group of three islands about twenty miles from the N.W. point of New Zealand, have the appearance of volcanic peaks, are barren in appearance and uninhabited. The Maoris visit them in canoes for fishing, I have been told. This spot is quite a favorite whaling ground, ships cruising for whales keeping from twenty to fifty miles off shore of them on three sides.

One morning, when the ship was some forty miles off shore, N.E. of the group, I started aloft just before sunrise to stand my masthead at the main. The sails that had been taken in at sunset the night before were hanging loose, as usual, for the men to set them when they came on deck, when all hands were called at daylight. Slowly climbing the rattlins, I arrived at the masthead about the time the topsails were being mastheaded, fore and main tacks boarded, and jibs hoisted, all at the same time. The rattle of blocks and braces and the men singing at their work raised such a din that a gun going off aloft would hardly have been heard, although only a light breeze was blowing and not a sea broke on the water.

After getting into the topgallant crosstrees with my feet, and my arms over the royal yard, I took a look around. The sun was just coming out of the line of the horizon, looking like a vast golden cheese. Hardly a cloud of much size could be seen in any part of the sky. I had been looking off the starboard bow, as we were on that tack. Turning my head towards the port, and at the same time casting my eyes down towards the lee bow, I saw something in the water that attracted my attention, but could not make out what it could be until the ship, slowly forging ahead through the water, brought the objects off the beam.

I first thought the things were whales, but taking a second look, could see no whale could be of such size, as the smallest of the three was at least one hundred and thirty feet in length. At the same time looking to the one in the middle, I saw such an immense length to this one, that if a man's hair ever did stand on end, mine approached that point. The other thing (or it, whatever it might have been) on the left side, was, judging by the length of the ship, to be of a length of about one hundred and seventy feet.

The middle and largest one was, at the lowest point I could possibly put his length, three hundred feet. One might say this is a big fish story. I do not care what any man may say. Although I might have been frightened at what I saw, I had not lost my head so much but I could use my poor judgment about their appearance as well as ever. I say, in speaking of the largest of these things, that from its looks the ship would have been more in its power to destroy than a whaleboat would be in the power of a fifteen-barrel whale, if size counted. They were heading away from the ship, which was slowly passing not over twenty feet clear of the tail of the middle one. This fact I noticed for the reason that I was afraid that if the ship should hit one, it might set the devil adrift in them and they, turning on her, might cause another missing ship to appear on the list.

It would be quite hard for me to describe the shape of things that appeared as they did; more so with the two smaller ones, as they did not seem to possess anywhere near as regular an outline to their forms as the large one, and were not very plainly outlined—more particularly the forward part. The head of the large one seemed to be nearly square across the end (something like the head of a sperm whale, but in no other way) and the width at the end was the widest part of its body. This, to the best of my ability in judging, was about thirty feet across. From its head, for three-quarters its length, would be about twenty feet wide. It then gained in width until it reached its tail, which was nearly as wide across as the end of its head, and nearly straight from end to end. The others were something the same shape, but not anywhere so well defined. By the time I had recovered from the astonishment these wonderful creatures had caused me, they had got abreast of the mizzin chains.

Hailing the deck as loud as I could, to make my voice heard above the din, it was a few minutes before the Mate got the decks still enough to hear what I had to say, and when he made out my bawling and pointing to the quarter, he jumped up on the poop deck and ran aft to see what I was pointing at. By this time the ship had passed them so far that he, not understanding properly what he had to look for, as these monsters of the deep were below the surface of the water, did not catch sight of them (and no doubt if he is alive today, he is still kicking himself he did not have the chance of a lifetime to see such a sight). Looking towards the water on the quarter for a minute, he yelled out, "What's that you see? Is it a sperm whale?" On my saying it was not, he told me to mind what I was at the masthead for and not stop all hands from work without good reason, and told the other officers to go on setting sail.

After all sail was set on the ship, braces hauled well taut, and the men had started to wash off the decks, the 2d Mate came up aloft to take a turn at the masthead until breakfast-time. When he had settled himself over the royal yard, with arms resting on it, he wanted to know: "What broke loose with you this morning? I hope you can explain it to the Mate's satisfaction, for he is wild to think you got him popping around on the poop deck, for what he don't know."

I soon told him what had passed the ship. He gave a grunt and wanted to know how my liver was. I told him the liver was all right, and that he was a bad man to treat a person who had been to so much trouble to get up such a show with so much disrespect as to grunt at him. "Still, the most you can get out of a hog is that, and I ought to be satisfied," was my answer, which he took pleasantly. In talking over the event, he felt very sorry that he could not have seen these strange creatures. "I would have very much liked someone else but myself to have seen them, also," said I.

When I was relieved and went on deck, the Captain met me as I got out of the main riggin', and wanted me to give him as near as possible a correct description of what I had seen. After my telling him all that I had seen, he said: "One hundred dollars would not have tempted me to miss the sight. After following the seas for over forty years, to have missed the only chance I ever had, in all that time, to see what must have been some of those apparently fabulous octopus, or squid, was too bad."

The other three boat-steerers kept up for a week a fire of what I called woodhouse wit, for the want of a better name, about the wonderful sight I had seen. However, I have not the least doubt in the world but that what I saw must have been those wonderful monsters of the deep. Many stories have been told of such submarine denizens, that have thrown their long arms over vessels and destroyed them; of boats dragged beneath the water and never seen more; but I have never sailed with any man who had seen anything of the kind. But I have no doubt, since I saw those things, that if they had arms or tentacles in proportion to their bodies, any of the stories told about them are correct. *

It is a well-known fact that certain members of the squid family have drowned many natives by throwing some of their arms around them and holding the victim to the bottom by others clasped around the rocks. The natives, in diving for fish, are caught sometimes this way. Taking into consideration the amount of strength displayed by that kind (which are not more than two or three feet long) and comparing it with the vast size of those things I saw, some idea can be formed of what they might do if they tried real hard.

I have seen a large sperm whale, when in his death struggles, vomit up misshapen pieces of these things almost if not quite as large as a whaleboat. This would float on the surface of the water after being ejected, so it was plainly

* "We now gazed at the most wondrous phenomenon which the secret seas have hitherto revealed to mankind. A vast pulpy mass, furlongs in length and breadth, of a glancing cream-color, lay floating on the water, innumerable long arms radiating from its center, and curling and twisting like a nest of anacondas, as if blindly to catch at any hapless object within reach. No perceptible face or front did it have; no conceivable token of either sensation or instinct; but undulated there on the billows, an unearthly, formless, chance-like apparition of life. . . . With a low sucking sound it slowly disappeared again.

"Whatever superstitions the sperm whalemen in general have connected with the sight of this object, certain it is, that a glimpse of it being so very unusual, that circumstance has gone far to invest it with portentousness. So rarely is it beheld, that though one and all of them declare it to be the largest animated thing in the ocean, yet very observed. There seemed not much body to it, the color on the outside a bluish gray, dotted with pink spots about the size of peas. Under this was milk-white flesh, or matter, that seemed to have no grain and afforded but little resistance to the hooked beaks of the albatross and molly-mocks that settled on the water around such dainty fare, tearing off pieces from the edges with noisy screeches and clatter.

WE CRUISED around the Three Kings for a month but saw no whales; and as we had to recruit for the northern season, we made sail for the Bay of Islands. Weather and winds being both in our favor, we soon ran in and anchored off the town of Russell. There we found two or three ships at anchor, none of which but had had better luck than us.

The *Mitchell* had taken oil enough to warrant his sailing for home. We felt somewhat downhearted to think we could not have taken oil enough to do likewise. Here was a ship with no better equipment than ours, cruising on the same grounds that we had; and he had taken in the same time more than twice as much oil, and now was fitting for home. Oh, how sweet sounded the word, to us that had to turn the old ship's head for another year in an opposite direction, in hopes we might get oil enough to make out a fair voyage!

There seemed no change in the town during our absence, except one more public house. This hardly appeared

few of them have any but the most vague ideas concerning its true nature and form; notwithstanding, they believe it to furnish the sperm whale his only food. For though other species of whales find their food above water, and may be seen by man in the act of feeding, the spermaceti whale obtains his whole food in unknown zones below the surface; and only by inference is it that anyone can tell of what, precisely, that food consists. At times, when closely pursued, he will disgorge what are supposed to be the detached arms of the squid; some of them thus exhibited exceeding twenty and thirty feet in length. They fancy that the monster to which these arms belong ordinarily clings by them to the bed of the ocean; and that the sperm whale, unlike other species, is supplied with teeth in order to attack and tear it."—*Moby-Dick,* LIX: "Squid."

needed, as there had seemed enough before, in a village of not more than two hundred people.

There lay close inshore of us a small vessel, sloop-rigged, commanded by a man by the name of Welch, and although there was not much about the looks of him or his craft to denote it, they bore the reputation of being the most noted on the coast of New Zealand. The notoriety consisted of smuggling but never getting caught. The sloop was fitted with a short jigger mast sticking straight up from her taffrail aft and a long bowsprit running straight out from the bows forward, which made her long and tapering mast look out of place between them. Her hull above water was anything but a thing of beauty, as it had the appearance of being modeled after a Dutch galiot; and to look for speed in a craft like that, one might think, would be like trying to find it in a green turtle on dry land. But sometimes appearances are deceitful, as I was told that no sailing craft on the coast could sail as fast as she.

We took on wood and water, painted ship, and then started to give liberty. The other ships' crews were having liberty at the same time, which caused a large number of men to be on shore each day; and it made the two or three policemen on duty in this small place almost crazy to keep any kind of order amongst so many wild sailors, some being full of fighting rum and the others more or less so. The three public houses drove a rattling trade during the days the men were on shore, as they had money to spend and but little chance to spend it for much else than drink of such a vile decoction that taken in quantities it rendered them quarrelsome. A great deal of fighting would be going on during the day amongst the crews of each ship, as, when in that state, things that had happened on board the ship and almost been forgotten would be brought up, and a fight would come off between the drunken fools; and other fools taking sides sometimes would bring into the fracas the rest of the watch.

But when two different ships' men got at it, the whole circus, with the band playing besides, would be in sight, for others from more ships would join in, and fighting would be the order of the day from one end of the beach to the other. This being some half-mile in extent and quite narrow, more fun to the square inch could be seen going on than a person would believe possible. Of course, it took but little time for the poor drunken fools to get so mixed up that they lost all idea of what they were fighting for; and striking out at any they saw before them, they made the performance the most ridiculous imaginable. The poor unfortunate policemen, at such times, would be dancing around on the outside like performing monkeys, brandishing their clubs, shouting "Order, in the Queen's name!" and getting tumbled heels over head by some of the crowd when they tried to drag off one of the number who had been laid sprawling on his back, too drunk to do more, or had been hit a little too hard.

I had a duty to perform at the close of some liberty days that to me was very unpleasant. Being one that did not care to mix up with those who, when on shore, could find no amusement except in drink, I had to get the men into the boat at sunset, as the liberty expired at that time, and bring them on board. During my efforts I would be almost driven wild in trying to induce them to get into the boat after getting them to her, but I never was struck by one of them, be he ever so drunk. One of my favorite ways to accomplish my will would be to get them in a body together, and, knowing their money had been all spent for an hour or so before and they could not get anything more to quench their thirst without it, I would say, "Now, boys, we will all take a drink and then go on board." Picking out one or two of the most sober ones, I would say, "Now, Bill," (or Tom, or whatever his name might be), "I expect you to see that all the boys follow me to the boat as soon as we have this drink, and don't let any stay behind." In this way I most always could get the men, if together, into the boat; but if they should be scattered in the different public houses, it often made it hard to get them together.

I never had any trouble with the women who kept the bars (as, in all English places where I have been, they do) but once; and I think this is the only time in my whole life that I ever spoke a bad word to woman or girl.

One of the public houses was kept by a woman that must have been at least fifty years of age, a hard-featured person, and not married, so I was told, to the man who attended the rough work about the establishment. Going into her house one afternoon for the men, as I had done before, I found them all there. Putting in practice my plan, to invite them to take a drink and then go on board, I stepped up to the bar, when she spoke to me, saying, "You never drink anything but ginger pop! Let me give you some nice raspberry wine."

"All right, I will try it for a change. Is it home made?" was my reply.

"Oh, yes, it's home made," she said. At the same time I saw her wink to her consort, who had a phiz on him like a man figurehead of some ship that seas had removed the paint from, and wind and sun had seamed in many places. But, thinking nothing of this, I took the glass containing the ruby-looking liquid from a bottle that she took from under the counter. Placing the same to my lips, I took a swallow and set the glass back.

She saw me place the glass back on the counter, and I suppose she noticed that I had drunk but little of it, for she came close to me and wanted to know why I did not finish the wine. "It won't hurt you," she snapped out. At the same time, her eyes fairly blazed with anger. I was mad, for by the taste of this stuff I thought I could detect some kind of drug; and the idea went through my mind that she wanted to so stupefy me with it that I would appear drunk, and not be the only boat-steerer in our ship that did not do so.

This being my thought, right or wrong, I did not stop to weigh words, but answered her by saying, "I think you are an old bawd of the most damnable type. What are you trying to celebrate, by putting stuff into that wine for?" She swore like a trooper that it was good wine and there was nothing in it to hurt anyone. "If that is so, let me see you drink," was my reply. She snatched the glass off the counter and emptied the contents on the floor, saying she would see any d—n Yankee in hell before she would drink his leavings. I told her, "That's all right—but I have not changed my opinion one little bit about there being some drug in that wine."

During the time the old cat and I had been having our little tiff, some of the men had crowded around me; and when she threw the liquor on the floor and made use of the expressions she did, some of them turned to me and asked if I wanted the whole thing piled out on the beach with the old bitch and her mummy of a partner on top. Most of them were so drunk they could not make out what the matter was, but could understand there was some trouble between the landlady and me. "Never mind, boys," I said, "we will go on board"; and they all followed me. But it was pretty hard work to keep them on the road to the boat, for first one and then another would get an idea that something had to be settled up in that house, and would want to turn back and do it.

A drunken man, to me, is an abomination; and although I have been able to handle them fairly well oftentimes, when obliged to, I have but little patience to put up with them when they are in the state that is called, "So drunk they do not know what they are doing." To me, this means so drunk that they do not care what they do; and no time do I believe the old saying true, except a person is so drunk he cannot walk or speak. That voyage instilled that idea in my mind, and it has ever remained there since.

The Captain of our ship was called one of the strictest men sailing out of the port of New Bedford. The whys and wherefores we will not discuss here, except in his treatment of men under the influence of liquor. His rule, and it had to strictly be obeyed, was to have an officer at the side with men enough to handle anyone coming on board at the close of a day's liberty so far under the influence of liquor that he needed help up the ship's side. A whip was rove at the main-yard for those so far gone that they could not crawl up the side with help. The men were handled as carefully as infants and landed on deck, then taken forward by their mates, undressed, and put in their bunks.

If one of the drunken men made the least noise or was quarrelsome, he was at once tumbled head over heels aft, and the irons clasped on him. If still he made noise, his knees were brought together and his hands forced over and below them, a pole or broom handle passed over one arm, under the crook of his knees, and over the other arm, rendering him perfectly helpless and looking something like a turkey when trussed. If he still continued to shout, or use bad language, a pump bolt was placed in his mouth and lashed with spunyarn around his head. In this condition he could breathe but could make no noise. In such cases it took but a short time to tame the most unruly, and they would be willing to do most anything to get relief from this torture. Men that had to suffer punishment were kept in irons until perfectly sober, and on being released, were told by the Captain to go forward, and see if when they came on board next time they kept quiet.

I know we had on that voyage some of the worst kind of

men when in liquor; and when we left the shore with them so drunk that they could not be got out of the boat except with the whip slung to the mainyard, their yells, shouting, and singing could be heard a mile away. But they would commence to quiet down the nearer we approached the ship, and when the boat shot alongside, a person on deck, not knowing to the contrary, would not believe a drunken man was within a mile. This, of course, was after a good number of them had been through the discipline.

Sometimes it made me laugh, as we neared the ship, when someone would start to sing a word or two of a song, but catch himself quickly, with perhaps a hiccough, and mutter to himself, "That old devil will hear me," and then keep quiet. Seeing what I state has caused me to form the opinion that a man, even when very drunk, can be made to mind if proper treatment is used.

One of my liberty days, while we were lying in port, I took a boat with three or four of the boys who also had liberty, and we started up one of the rivers that empty into the bay. Pulling and using the sail for about five miles, we entered the mouth of the river. After pulling up about three miles, between marshy banks covered with dense growths of wild flax that stood six to eight feet in height and hid completely the shore and all except the high land in the distance, it became monotonous and we had about made up our minds to turn the boat around.

On rounding the next turn, though, a view of such loveliness opened out that we felt well repaid for our trip if we saw nothing else. I told the boys to peak the oars, and we let the boat lay and took a look. In the distance quite a high range of hills could be seen, extending to the right until lost to view in the tops of a forest of trees that covered rolling land as far as the eye could reach. An opening on the right bank of the river, clear of trees, showed a tract of land rising gently as it ran back until it met the forest trees that stood so thickly that their light-colored trunks, standing like huge posts, seemed to forbid further approach to the bright green grass that covered this opening.

This oasis was perhaps five or six hundred acres in area, of a semicircular shape. Two or three settlers' houses, with the outbuildings painted or whitewashed, showed in fine contrast to the surroundings. Cattle and sheep could be seen feeding here and there in this tract of land that had, to a certain extent, the appearance of an immense lawn. On the left bank of the river, open land could be seen amongst clumps of trees that gave the look of an immense park. The delusion was rendered more striking as no underbrush could be seen, and the grass covered the ground in all places clear of the trees.

Some native huts could be seen close to the river banks, in front of which some men and women stood waving for us to come on shore. Some of the boys wanted to land, but I told them we would pull up along the river a bit farther and when we returned might stop, as I was pretty well aware that if we landed now it would be hard work for me to get them much farther up the river that day.

Pulling ahead for a short distance along the river bank, I thought I could see some peach trees and other fruit trees a short distance from the bank of the river, between us and a fine-looking house. Steering towards the shore, I saw a pair of steps that had been made in the steep bank that the river had, along here. We shipped in the oars as we ran alongside it, and one man jumped out with the warp. I told the boys to wait while I ran up to the house to see if we could beg, buy or borrow some fruit.

Approaching the house, I could see a little stir going on, as though they had been watching us. I stepped up to the door and knocked. An elderly man opened it. He had a bright rosy look to his face, and a pleasant smile came over it when I told him we were a party from one of the ships that lay at anchor in the bay and were spending our liberty day by taking a trip up this river; and that seeing some fruit growing in his orchard, I had landed to find out if we could purchase a little. By this he stepped out from the door with a hearty, "Ah, man, but you cannot buy any of my fruit—not a hap'nny worth, my word you can't! But you and your men can take as much of the blooming stuff as you jolly well mind, my oath on it."

Going down to the boat, I told the men to come up, as we had permission to get some fruit. When we got up to the fruit trees, the old Englishman, his wife, who was as jolly-looking as he and broader across the beam than she was long, and a girl of about sixteen who must have been a daughter, as her dimensions corresponded to the mother in more ways than one, were waiting to show us the best trees and help us pick the choicest kinds. As the young lady had on quite a short dress and no shoes or stockings, a fair share of her lower works showed when stooping over to pick up some peach more fine than others, making those having a stern view inclined to run away from the ship and settle down on some of the land near this place. At least, they said so afterwards.

In talking with this gentleman, while eating his fruit, he told me that he had been living here for a number of years; and in answer to my question, if he had ever had trouble with the natives, he told me that he had not, but many other settlers had. The natives, he said, were very bold fighters and used much strategy in their battles; and the military had found out to their sorrow, quite a number of times, that they were a foe worthy of English soldiers' contending against. One instance that he told me about caused me to think he could not be far out of his reckoning, after all.

"It happened," he said, "after the last outbreak of the Maoris, who had been fighting with the soldiers and colonists for some time, that a large body of the Maoris had collected on a hill well-chosen for fortification, in the manner they do such things, by driving heavy stakes in the ground and cutting the ends out of the ground to sharp points. These are placed so close together that bullets at much of a range will not penetrate. Three lines of these had been set into position, and behind the first row, on the side of approach, the Maoris, well-armed, lay waiting the advance of a body of troops and colonists.

"The Maoris, behind this breastwork had places made through the stakes large enough to place the muzzles of their guns, and sighting them, these holes were so low down that they lay flat on the ground. Troops advancing up the hill and firing, scarce a bullet from their guns would strike except above them. As soon as the Maoris could make a shot tell, they commenced to fire, and many a poor fellow bit the dust before the first row of stakes was carried.

"The Maoris by this time had all retreated behind the second row, and the same game was carried on until they had to retreat behind the third and last. In the meantime, redcoats were lying dead in all directions.

"The soldiers, maddened to desperation, carried the last stronghold and found no one in it alive or dead except an old Maori woman. A hole in the ground, connecting with a tunnel that opened into a gulch, told the tale of the disappearance.

"Over five hundred soldiers had been killed and many wounded, but as far as could be seen not one Maori had been slain," so he informed me. He told me that there had been no fighting going on for quite a bit on the North Island, but it was not all quiet on the South.

Eating all the fruit we could, and taking quite a lot besides that he kindly gave us, we shoved off, after thanking him, his good wife, and daughter for the favor shown us.

When we got abreast of the place where we saw the Maoris on going up the river, the boys wanted to land so much that

I ran the boat alongside the bank, and we went on shore. Two or three men and five or six women and girls met us as we went towards one of the five or six houses. These women and girls ranged in age from fourteen to forty, and some of the younger ones were quite fine-looking, even if their skins were brown-yellow. One of the girls, by her looks, had quite an amount of white blood in her, as she was light enough to show quite red cheeks, and freckles showed on her hands and arms. She had black curls hanging quite low down her neck and was dressed partly in European style, which gave her the best appearance of any I saw. She was about sixteen or eighteen years old, I should think.

There was one thing I noticed in the natives of New Zealand: that they had more the look and manner of the Sioux, Winnebagoes, or Crow Indians, than they did any Kanakas of the South Sea Islands.

We stayed around there for a couple of hours and I bought about 30 lbs. of honey, in the comb, and two or three kits (or flaxen baskets) of peaches. For honey and peaches I think this part of the world beats anything I have ever seen. Cheap is no name for these things. The honey and peaches only cost me a crown, $1.25.

Pulling and sailing, we got to the ship just after dark, having spent one of the pleasantest day's liberty of any I had while lying at anchor in the Bay of Islands. The honey I brought on board was much enjoyed by the other boatsteerers, as well as by yours truly.

THE CREWS of all the ships lying there had their liberty brought to an end by one of the most desperate fights that had ever happened between foreigners on that beach. One Sunday most of the crews of the ships were ashore, leaving but a few men in each ship. Quite a few settlers had come from different places adjoining the bay to pass the Sunday away in the town. Also, some of the soldiers from a station some four or five miles up towards the head of the bay had come to pass away a day's leave of absence. No doubt but in round numbers there must have been at least 150 or 200 men on the beach that day from the outside. I did not care to go on shore, but the other three boat-steerers had gone. The 2d & 3d Mates were on board; the Mate and 4th Mate had gone to visit some acquaintances a short distance away from the town. So the old ship was pretty near deserted, as there were only left three or four of the crew forward, enough to man the boat and bring off the men at sunset. They, with the cook and steward, could not be spared.

About three o'clock I heard one of the men say, "There seems to be trouble on shore." Looking towards one of the public houses (they all three stood out plain in view, on a ridge of land back of where the boats landed on the beach, although some distance apart), I saw quite a crowd of men in front of it; and the next minute I saw a red coat with a man in it roll end over end out of the crowd and halfway down the beach towards the water. He lay there for a few moments, then scrambled to his feet, made towards the men and soon was lost amongst swaying bodies and flying arms. Stepping to the cabin companionway, I called to the officers who were below, telling them that wigs were being scattered on the green ashore.

By the time they got on deck, sailors, soldiers and landsmen could be seen running on the beach and piling helter-skelter out of the other two public houses and making for the center of attraction, where by this time men could be seen tumbling over in all shapes. Some would soon right themselves up; others would lie stretched out full length, back down; some on their sides, knees, and in all kinds of positions. Now and then one could be seen "carrying off his eyes with both hands," as he slowly crawled to one side. In this way numbers dropped out, but it did not diminish the crowd that were improving the shining Sunday hours to see who could hit the hardest, as men were constantly arriving and not hesitating, when joining the crowd, to present their letter of introduction by running their fist against the first head that they came to.

Looking from the ship, we had a top seat above the circus ring, and although it did not have a pleasing look, by any manner of means, blest if I could leave looking at what was going on, but seemed to find such excitement about it that I could hardly look away. There seemed to be an extra amount of redcoats stretched out along the beach, as though the body acted on that department with more vigor than on others; and by the time the sportive lambs had fought to the limit of one end of the beach and began to work back again, hardly a redcoat could be seen on his feet.

The entertainment continued back along, over the fallen who had not strength left to crawl out of the way or were too full of bad rum and knocks to care, scattering, as the body moved, more victims to help fill the number that were holding the sand down. Twice up and down the battlefield moved the hosts, until perhaps half the number were laid hors de combat and the balance somewhat damaged, more or less, in the upper works. The fight continued for at least two hours.

But like everything else, an end had to come. A squad of soldiers arrived that had been sent for by the police, as they had no show to quell the disturbance; and a show of fixed bayonets and a few knocked down with the butts of guns soon had the field clear of combatants. The sick soldiers were marched off under guard to their barracks and the police arrested every sailor they could lay hands on. By sunset the beach around the public houses looked like a deserted village.

Our men on shore had been locked up in an old building, as the jail could not hold more than half that had been arrested. Both this and the jail were wooden buildings, and when our fellows sobered up, they broke out of where they were in confinement and went over to where the others were in jail, and helped them break out of that. No one interfered, and it was not known if it were because of fear, or not knowing about it.

Shoving some small boats off the beach, some of the men went off in them and reported to each ship that the men were on the beach and wanted boats sent so they could get on board. Boats were sent, and long before daylight every man was on board his ship; but the next morning some of them had heads on them that looked as though a horse had trod on them with shoes on.

The next afternoon an officer came on board to find out the names of those who had been on shore the day before; but strange to say, he could not find out that any of our men had been out of the ship during the day, and he had to leave without making any arrests. The affair assumed a serious look as the officer, before he left, informed the Mate that two or three men had been killed and a number very dangerously hurt by stabs and cuts.

As our ship was ready to sail, we got under way the next morning and went to sea. We never heard afterwards who did the stabbing; but the boys who had been in the row thought most likely it was done by the soldiers, as some of them had, in the sheaths at their sides, bayonets.

AFTER getting clear of the land and stowing the anchors on the bows, we shaped our course for Sunday Island, to take a short cruise around there before we went North for another cruise around the Line and the Fijis. Thus we spent some six weeks, and saw whales three or four times, but came no nearer getting any than striking two and having the irons draw from both.

A few days before we left that ground for the northern cruise, we sent two boats for fish amongst five or six small islets laying off shore on the east side of Sunday Island and distant from it from two to eight miles, one of which is large enough to have quite a growth of trees and shrubbery. On the North side of this can be found a fine little boat harbor, but a person unacquainted would be apt to pull by without seeing it. After we had caught quite a number of fine fish, we pulled into the boat harbor and had our lunch, then scattered over the island hunting for eggs and paroquets. These last-named are quite tame and often can be caught in the hand. Some of the boys that day caught three or four to take on board ship, and in a few days they became tame and were quite pets.

The eggs to be had there are very like in shape and taste to hen's eggs. The birds that lay them are a species of gull, called by the whalemen mutton birds, for what reason I know not; for the old birds have about as much flavor of mutton as a carrion crow has. The eggs and young are found in holes in the cliffs, two in a nest, lying on the bare rock without the first sign of a nest. The young have a most wonderful appearance before they are old enough to leave the nest, covered as they are with a soft puffy down, of a bluish color, with just the end of a bill sticking out from a round ball like fine cotton batting, larger than the body of its mother. Seen for the first time these would astonish anyone as much as it did me the first time I reached into a hole and pulled one out. They are one mass of fat, not a streak of lean in their whole bodies, and they can be eaten by boiling; but one does not hanker after them. They are used as a substitute for meat by splitting them open on the back, taking the entrails out and the down off, and then packing them in salt for a time.

We spent some three or four hours hunting for eggs, and returned to the ship with three or four bushel. All hands had a big blow-out on fresh fish and eggs. After cleaning the fish and washing off decks, supper was called. When that was through, all sail was made, whole watches were set, and we steered a course towards the North that would about strike the S.E. point of the Fijis.

Close Chances

NOTHING of importance took place on our run towards our cruising ground. The watch on deck was kept busy as usual in mending sails, repairing service on the riggin', taking off scotchmen (pieces of oak strips hollowed on the inside to fit the riggin' they were seized to, for preventing the chafe of the spars or ropes that they came in contact with), putting a coat of tar under them and replacing them with new ones where needed. Rattlins to straighten up and new ones to be seized on, making spunyarn, scraping masts, cleaning spades; these and many other things kept the men busy all the hours between seven A.M. until four P.M., day after day the voyage through, when the weather permitted.

We made a fine run of about two weeks and took in sail at sunset about twenty mile S.E. of the most southern island of the Fiji Group. The name given it by the whalemen is Farewell Island, and to all appearances it is uninhabited. Cruising around here for a week or more, one day we were aroused from our dull monotony of ship duty by the pleasing sound of the cry from the masthead: "There she blows! T-h-e-r-e s-h-e b-l-o-w-s!"

After the answer to the Captain's question of "Where away?" had been given, the mainsail was hauled up and the mainyard hauled aback. The spouts of a fine school of whales could be seen broad on our lee beam, about two miles off; and by the way the bushy spouts were slowly sent into the air, we could see that they were very quiet and heading to the leeward. The lines were hurried into the boats, after all hands had been called. Down went the four boats into the water, up went the sails and all headed to the leeward, with sheets well off, the boats' heads pointing towards the spot where the spouts had been last seen, as the whales had gone down during the time we were lowering away.

The sea was quite smooth, the wind blowing a good whaling breeze that sent the boats through the water with a roll of foam sparkling from each bow, showing in its whiteness a beautiful contrast to the azure blueness of the sea. The sun was shining brightly from a clear sky that rivaled in blueness the ocean over which the boats were bounding with a motion that seemed to partake of the spirit of the crews in them.

As the boats under sail would have plenty of time to reach the spot where the whales went down by the time they would make their appearance, no oars or paddles were used, so we all sat in the boat, as the time approached for the whales to break water, looking to see the first spout. I suppose one officer did not like to bring his boat to the wind before the other, and the result was that all at once the whales came up all around us, but unfortunately none within dart of our irons. We had sailed a little too far over them; and by the way they sent their spouts into the air with a noisy roar, they proclaimed the news to our disappointed ears that they were badly gallied. Each boat pointed for a whale, hoping to get within dart before they recovered from their fright and started off in a body to the windward, which, as

I have said, most always is the case when whales are gallied.

The whale we started for was a cow and she had with her a calf that would perhaps make about five barrels of oil. With the sail hard full and the four oars pulling (as the 2d Mate had told the others to pull but for me to take the first iron in hand and try, if possible, to hit a whale) the boat went flying through the water pointing towards the cow, but with an instinct that these animals possess in a most wonderful degree, she dove into the depths just out of reach of an iron. The calf was a little slower in its movements, so when the boat shot over the spot where they settled, I saw the shape of the little beggar under water, but quite a distance below the surface. Now I did an act that only success authorizes, which was to dart my irons one after the other at the whale, neither of which hit him for the reason no doubt that he was too deep, or that the boat was moving so rapidly that my aim was faulty.

My not hitting this whale was, for a man in my position, a pretty serious affair, for I had darted at the whale without being told to do so by my officer. If he had told me to dart and I had missed, the case would have stood different, although the results would have been the same. When I darted at the whale, the officer could not see him; and well knowing that the chance was very slim of my hitting him, I thought that perhaps one of the irons might enter his body far enough to hold. But there was not time for me to tell him about what I was going to dart at, as before he could understand about it and give me permission to dart, the whale would have been out of sight and the boat a long way past where the whales went down. This went through my mind and caused me to take a chance that, in failing to strike, might cause my being disrated and forever bar my chance of steering another boat in this ship. For the rule laid out by our Captain, in regard to that, was, no boat-steerer can miss but one whale and have a second chance to miss another.

If this took place with me, all future hopes of rising in the profession of a whaleman were doomed, and I would have to choose some other vocation. So, when the 2d Mate found out that the irons were not into the whale and told me in a cross tone to haul them in, it was with anything but happy feelings that I obeyed the order.

I was feeling very downhearted to think I had been so unfortunate as not to be able to hit the little devil (and what might be the consequences of it) when the 2d Mate commenced to rip and roar about my missing the whale. After a little time I could not stand any more of his abuse (knowing, as I did, that what I had done was the best that could have been done under the circumstances), so I quietly asked him what was the matter with him; but I was hopping mad just the same, and reckless of whatever might come.

"You ask me what is the matter with me?" was his reply.

"Yes, that's what I said, and furthermore, what do you mean by saying that I missed a whale? Please tell me that now, if you can stop your abuse long enough to get your mouth clean enough for you to do so."

By this time he was running over with passion, and swore he would come forward where I was and throw me overboard if I did not shut up.

"All right," I said, "but before I shut entirely up I will ask you, how do you make out I missed any whale? This I want you to answer before I shut up one little bit. You started this thing, I will keep it going until you give me an answer to that question, let it be in board or overboard, it matters not to me. Do you mind that, now? I suppose you mean what you say. I know I do."

For a minute or so he said not a word, but if looks could have riddled me, I would have been as full of holes as the cook's old skimmer hanging on a nail in his galley. Well, I did my best to look at him with as much as would punch holes.

Then he yelled out, "You missed the whale! You darted and your irons never went in. What do you call that?"

"What do I call what? Do you know what I darted at, or not?" was my quiet reply.

This staggered him, as I knew he could not possibly have seen either of the whales after they had settled. He flopped over with passion and I did not know but what he meant to come forward where I was, as he made some feint to peak his steering oar, but changed his mind.

"Look here, young fellow," he said, "if you did not dart at a whale, what did you dart your irons for?"

"Now, as for the first time you have asked a leading question and are so very pleasant, I will impart for your information in the most kind, condescending, humble manner, and with much affectionate solicitude for your future welfare, in the vain hope of relieving your poor benighted mind, that I did the best I was able, with the two irons I darted, to fill the hole in the water that those two whales made when they went down."

For two or three seconds everything was quiet. After I had finished, I did not know from his looks but what my time had come and I had best make a break over the bow before he slaughtered me. Then came a roar of laughter from the other four men in the boat that might have been heard half a mile off.

The only reply he made was, "I will attend to this matter when we get on board the ship." The other boys burst out in a quiet chuckle as we pulled slowly to the windward, after he, the 2d Mate, had ordered the sail rolled up and the oars manned. Since the whales had disappeared, we followed the other boats towards the ship.

After hoisting up the boats, unbending the irons, and taking the lines out of the boats and placing them on their racks, we boat-steerers assembled in the waist. To the question asked by them of me, "What was the matter?" as they had seen me dart, I gave them no other reply than, "You will see a Barnum show, perhaps, before long."

The Captain had been walking the weather side of the quarter-deck during the time the boats were being put in place. He was looking mad as a sailor when somebody has stolen his plum duff. When the mainyard had been filled and maintack boarded, he called the officers' attention to the fact that owing to their bad judgment, a school of whales had most foolishly been driven away without taking any. He ripped and swore like a pirate. Go to it, old fellow, thinks I. The more you get clear of, that way, the better chance for me when my turn comes.

I did not have to wait long, for after the 2d Mate had been talking a few minutes with him (what was said, none of us in the waist could hear), I saw the old fellow turn quickly and with a fearful scowl on his face look towards where I was standing with the other boat-steerers.

This was the first black look he had given me in all the months I had sailed under him. I felt it most keenly, for although he bore the name of being one of the most stern and strict captains out of New Bedford, I liked him, as so far he never had found fault with me or my work. So when he sternly called me to where he stood with the other officers, I cared nothing for missing the whale in comparison to having him so angry towards me.

The first word he said to me caused me to think my doom was sealed. The thought of my poor mother, who had begged me almost on her knees not to follow the sea; now after nearly seven years in the service, to return when the ship did, a broken boat-steerer; the thought of the jeers that would meet me when walking the streets of New Bedford, from cousins and old schoolmates who had advanced in this profession: all this was flying through my mind as I stood waiting to hear the charges the 2d Mate had brought.

By the time he had finished, I could see he had not understood the case or had been wrongly informed. The 2d Mate had not said anything more about the little conversation we had had than that I was impudent to him after missing the whale. He, the unreasonable man, had informed the Captain that after I had darted my irons, the whale I had darted them at came to the surface; and if I had not darted when I did, there would have been no trouble of striking one, but the irons' being overboard was the cause of my not striking one when they came up.

The Captain was raving. A white heat was no name for it. He wanted to know what I meant by darting the irons as I had, and said, "No boat-steerer on board this ship can have a chance to play a trick, like that you have been guilty of, twice. But before I condemn you, I want a reason for your unwhaleman-like action."

During the time he had been railing and storming up and down the quarter-deck, at times approaching me with a look on his face that the Devil might have envied, I kept getting more mad myself, to see that I was so unjustly treated. So, when he asked me for the reason of my darting when I did, I burst out with the reply, "What is the use of my saying anything when your 2d Officer stands there and tells a most damnable bare-faced lie?"

Well, as I told the other boat-steerers it would, a show did take place when I rounded out the word "lie." The 2d Mate fairly shrieked, "What do you mean by calling me a liar!" He jumped up and down, swearing he would break every bone in my body and saying a whole lot of other stuff. The 1st Mate muttered over a lot about a good place for me would be in the mizzin riggin' with a dozen cuts with the cat before I was sent forward. The 3d Mate had nothing to say. I knew how he felt towards me, for he had been for a long time dissatisfied with his boat-steerer and had wanted the 2d Mate to exchange, but the 2d Mate would not. The 4th Mate did not seem to care how it went, but, like the boat-steerers, who were hanging around the fife rail of the main-mast, liked the fun of the thing.

The Captain, who I expected would play the deuce, stood still a minute, and a quiet look of astonishment came on his face as he called the 1st and 2d Mates by name and requested silence. Then, turning to me, he advanced close towards where I stood and in a calm voice, that had a ring in it to be more feared than his loud words of anger, said:

"What do you mean, sir, by calling one of the officers of this ship a liar?"

"What I mean by calling your 2d Mate a liar is that no whale came up anywhere near the boat, after I had darted my irons, that could have been reached with a rifle," was my reply. This seemed to startle the Captain; and as the 2d Mate attempted to say something, the Old Man told him to keep quiet, that he would finish with me first, as by what I said there was no doubt a lie somewhere and he would come to that later on. In the meantime, he ordered me to answer his question, why I had darted my irons without having been ordered by the officer to do so?

I had recovered myself somewhat, and replied that I was sorry to use such a word to an officer of this ship, as I had, and would state the facts of my reason in darting. When the 2d Mate told me to take my iron in hand, after the whales came up gallied, he said also for me to try and see if I could hit a whale. "That means, if I understand it right, for any chance that might offer to strike a whale, to try and do it, if I thought it possible." I then stated to him how and why I came to dart; at the same time informing him that the 2d Mate could no more have seen the whale I darted at than he could himself, that was on board the ship, so he could not judge whether I had missed or whether he was too deep for the irons to reach. And as regards this, I could not say the whale was near enough myself. "All I know is, the irons did not hit him," said I, "which I am sorry for. But if I am broken, you will not say that it was caused because I was afraid of the whale.

"Now, as regards the whales coming up, so that if my irons had been in the boat, one might have been struck, I did not see any but those that came up so far off that we gave up the chase after a short time and came on board. If any one of the men that were in our boat can say that they saw any whale within striking distance, then I am ready to suffer for my mistake in any way you see fit."

When I finished, the Captain turned and walked aft in a slow manner. Coming to the cabin gangway, he went below. In a few minutes the cabin boy came up and told the 2d Mate the Captain wanted him. He went into the cabin, and for a half an hour or more they were together.

The cabin boy said that the Captain gave the 2d Mate a big tongue-lashing for his conduct, and told him that I was not to blame enough to warrant my being broken.

The 3d Mate went to the Captain afterwards and told him that he wanted me to steer him, but the Captain told him that he could not make the change unless the 2d Mate was willing, as it would not be right; but if they could agree between them, he was willing. I wanted to make the change but the 2d Mate was not willing, after all. It was some months, though, before there was a good feeling between us. I can hardly tell what made him do such an insane thing as he did, except he was so wild with me that he took that course to blast me forever as a whaleman.

There is no doubt but what sometimes good men are disrated, unjustly, on board a whale ship. One instance happened of that kind, or would have happened, which I may here describe, on board a ship that I later was officer on, of a first-class boat-steerer being broken, who never could have become anything above a foremast hand on any whale ships afterwards, except by my helping him to prove he was a good boat-steerer. But afterwards he rose to as high a rank as an officer on board other whale ships as his education would permit. As I always have a feeling of pleasure when I think about this affair, it seems that others would likewise, so I will mention the occurrence.

It was my first voyage as an officer. I had been very successful in getting whales to my boat and under such conditions that the Captain thought I was all right, so, in many ways that would not affect the discipline of the ship, he made it manifest, and gave me my way in most matters that did not interfere with the other officers. I was 3d Officer. The Mate was a stern, disagreeable man of about forty, who seldom spoke a pleasant word to anyone, not even the Captain, and the manner of his treatment to the men often made my blood run wild with anger; but as an officer under him, I was powerless to say anything. He was one of the most perfect men in care of the ship and riggin' that I ever sailed with, but he did not care to get any nearer a whale than he was obliged to, and I knew it.

My boat-steerer got infatuated with a pretty Spanish girl in a Spanish port we stopped at, and he ran away. The chance offered of shipping an experienced boat-steerer in his place; a man who was capable of heading a boat, and a first-class sailor. So the Captain paid him a good round lay, as he said to me that he wanted me to have as good a boat-steerer as could be got. Of course, this just suited me.

The next day after we left port where we shipped the new man, the Captain called me into his cabin and informed me that the Mate had told him that he would not allow his boat-steerer to go in his boat; that he could not strike a whale, and if he expected him to get whales, he must have a different man to strike them, as the one he had was good for nothing in that respect.

"What are you going to do about it?" was my question.

The Old Man, I could see, was bothered; and he said that he hardly liked to take away the boat-steerer he had shipped for me, and let me pick out one from amongst the men, without my consent, nor did he like to break Crocker in such a manner. It was plain to see how much he was annoyed by his Mate's cussedness. When he stopped talking, I asked him if he thought that Crocker was not fit for a boat-steerer? The reply he made was that, as far as he knew, he was, but the Mate, and not him, decided he was not; and as the case stood he could not compel the Mate to retain him.

"Well, Captain, a short rope is soon unrove. I am willing to give up the boat-steerer you shipped for me and let the Mate have him, providen I may have Crocker to steer me," was my reply.

Rising out of his chair suddenly, he spoke quickly, and with feeling, saying, "You want Crocker to steer you, when the Mate refuses him? Certainly, I am willing you should have him, but I hope you are not letting sentiment take the place of judgment. That and catching a whale are wide apart. What we all do should be for the benefit of the voyage—our feelings do not count for anything."

"No, sir," was my answer; "I have the interest of this voyage in my mind as much as any common man on board this ship, and if, on proper trial, a man fails to do his duty, he will find me as hard and as just as man can be, even if it should be my own brother. As for Crocker, I believe that, get him near enough to a whale, he will do his whole duty and strike him, for he is no more afraid of a whale than I am and perhaps not as much. And although it may be treason to say it, I must under the circumstances do so—that if the Mate had placed his boat nearer some of the whales he has told Crocker to dart at, he would have struck them all right. He, I think, needs, like myself, a close chance, as he cannot dart an iron far. But you may trust me not to allow any loss to the ship in trying to prove what I say."

After a little further conversation, in which he expressed a good feeling towards me, we went on deck. The change was effected and the Mate got one of the best all-round boat-steerers I ever saw. Crocker, as the voyage went on, proved he could strike whales all right, as he had the pleasure (as well as myself) of bringing two whales to the ship, to the other three boats' one, all the rest of the voyage.

I must say a few more words about Crocker before we continue our cruise. In the first place, he was the most pleasant in manner of any man on board ship, and never could do enough for me; not much of a sailor, but for daring, when alongside a whale, he was the coolest man I ever was in a boat with. A whale might strike or fight with his head or jaw, it mattered not to Crocker. I have put him up to a whale and seen him put his irons into it when I do not know but I, being placed in a like position, would have hesitated; but he never faltered and would drive them home with as much calmness as a woman would send a churn dash into the churn when making butter. The satisfaction he would exhibit on our getting a whale used to do me as much good to see as it would be to get the whale.

Many times, when we had long pulls, he would so inspire the boat's crew with songs and words that we most always came out ahead and would be able, if the chance offered, to strike a whale by so doing. So I as well as he profited by the change. The Mate's boat-steerer complained to the others of the distance he kept from a whale, and they told Crocker that he was often wild that he was taken out of my boat.

Well, I guess you, who read these reminiscences, will tire if I do not get back on my course again.

Let's see; just before I yawed out of my course, the investigation of my missing, or not missing, a whale was about closed. The Captain had a talk with the 2d Mate in the cabin; and shortly after, the officer came on deck. The Captain come out of the cabin also. He took a turn or two up and down the quarter-deck, then stopped abreast of the mainmast, and called me to him from the waist, where I had gone amongst the other boat-steerers, just after he went below. After I had come close to him, he sternly said, "Young man, you had better be careful how you carry sail on board this ship, and not cut the corner of your flukes too far out of water! You understand what I mean, sir?"

I told him I did (as I supposed he had been told by the 2d Mate of our little side-affair in the boat). "Now you go on with your duties, as before, and look out you make no mistakes when you are in the head of a boat. And remember, you are nothing but a boat-steerer—and pay strict attention to what the officer that heads your boat says. And mark well, no back talk. That's all! Go to your duty and let me hear no more complaints."

When I got below with the other boat-steerers, they all thought I was, as they expressed it, a h—l of a fellow.

We continued the cruise along the reefs around the Fiji Islands and took two or three whales; but nothing happened out of the common in their capture except that a little calf belonging to one of the whales, taking our boat for his dead mother that lay floating on the water a short distance off, bumped its little head through the bottom of our boat and set us in it afloat. But by stuffing some of our clothes into the hole he made, we managed to reach the ship with two oars, the other three men bailing.

The calf went around one of the other boats and an iron was thrown into it, and after a thrust or two with the lance the poor little thing soon lay floating dead on the water no great distance from its mother, which could not have given birth to it more than a day or so before. The little thing was hoisted in whole by one tackle. The little whale gave us boat-steerers a fine lesson in regard to the anatomy of its species, and was to me a great help in reaching the life of many a sperm whale later on. He made about two barrels of oil.

One day, when some five or six miles off shore from the

reefs, and abreast of a passage leading through it that would give communication with one of the large islands that the reef here guarded, a large double war canoe hove in sight, heading for the opening in the reef. The sea was smooth and the wind was blowing a gentle S.E. trade that just filled our sails and sent the old ship's bows through the water with only force enough to cause a ribbon of sparkling foam to roll out each side and dance in merry bubbling streaks along the bright copper on its way aft, as the ship lazily told off three or four knots an hour. We were close-hauled on the offshore tack with our starboard tacks aboard. The canoe was heading in shore with his sheets eased off, and as he was dead ahead, if neither changed course we would pass very near together. As neither changed, we ran by each other so close that we almost touched sides. As we passed her, on a platform amidships of her could be plainly seen the big part of a man's leg that had been partly roasted, from which pieces had been hacked or torn. This display might have been for our benefit —or it may have been the meat locker. There was no doubt but that these fellows had been off to some of the other islands, and had captured some poor unfortunates and had a good roast, and were now returning home to exhibit the heads of those they had dined on. Anyway, we who saw it almost turned sick at the ghastly sight, and curses were showered at the yelling black devils as they went by.

The canoe, like others of its kind, would no doubt hold at least two or three hundred warriors. This one had them packed all over it, as close as they could sit or stand, so it appeared. How far they had been, there was no telling, but if they had traveled far, I should think they would be glad to get on shore and stretch their legs.

It is a well-known fact that those canoes can go long distances at sea and are capable of standing heavy weather. They sail well, and, the way they handle them, are very seaworthy. The sails they spread are immense. This one had almost as much spread on it as our main course, and like all others on the Pacific, they were made of mats.

We had a fine exhibition of the sailing and sea-going qualities of these huge affairs, one day when we were pitching lee-bow under, off the reefs. A strong breeze was blowing, and the old ship was heading towards the point where some lee might offer a better chance for lowering for a whale, if one should be sighted, than it did in the open water offshore. The ship had only double-reefed topsails, fore and main courses, jib and spanker set, and, as we were close hauled on the wind, it was sail enough.

One of the officers who was at the masthead sang out, "A large canoe in sight, on the lee quarter, heading the same as the ship." In two hours the canoe had overhauled us. She was abeam and not over a half-mile to the leeward. This gave us a fine show to see her as she dashed through the seas, as much under them as she did over them. Seeming not to heel over, she sent wreaths of foaming water by her sides, somewhat as a sled runner might dash light snow in running through it.

There was quite a body of black devils huddled amidships. How many would be hard to say, as they seemed too uncomfortable with the spray flying over them to hardly look at us, much less rise up. But it was a pretty sight to see the thing dash by us with the mat sail swelled out to its fullest extent, the sheet and halyards taut as bars of iron, sending the long covered ends of the canoes half their length into seas that they hardly raised to ride over, or cared for. On she went, gaining to the windward and ahead of us so much that when we shot into smoother water, under the lee of the reefs some five or six miles to the windward, two or three hours after she passed us, the canoe was as far ahead as could be seen, bearing two points on the weather bow.

Our ship was a good sailor, and weatherly in heavy weather. During the time the canoe was in sight, the wind was steady. Our progress through the water could have been no less than six or seven knots. That craft must have sailed nine or ten knots and gone to the windward besides.

Nothing out of the common course of the dull monotony of a whaleman's life took place the rest of the cruise, except that we nearly ran ashore on a sunken reef that was not laid down on our chart, some eight or ten miles off the easternmost island of the group. But, as we kept at least six inches between the top of it and the bottom of our keel, no harm came of it, though the crew had a lively time for a spell to get the ship clear.

As it was getting time to replenish our water and take more wood on board, besides giving the men a run on shore to keep the scurvy out of their bones, the ship's head was pointed towards the island of White Heads, so called by the sailors, and marked on the chart by the name of Rotumah. It is situated in about the Lat. of 12° South, Long. 177° East of Greenwich, about 250 miles North of the Fiji Group, and is of a purely volcanic formation, having in many places extinct craters, none of which are of much size, and showing, by the heavy growth of trees and decomposed vegetation that has caused a deep soil in them, that many years must have elapsed since they were in action.

The island, I should judge, was about twenty-five miles in circumference, nearly round, with a reef that encircled it a short distance from the shore, but here and there, from some unknown reason, the fine white beach made into the water that showed no signs of rocks, reefs or shoals; a bathing spot for the gods. The white sand beach, extending entirely around the island, showed, in fine contrast to the bright green border of shrubbery that hung over the upper part, as a white ribbon might on bunches of evergreen.

The second day after leaving the Fijis, a short time before sunset, the boat-steerer at the main (no officer has a masthead after supper, which is at five P.M., when no whaling interferes) sang out, "Land ho, right ahead!" At sunset the light sails were furled, the fore and main course hauled up, and the ship hauled to the wind with the main topsail to the mast, on the starboard tack heading to the N.E., the wind blowing a gentle gale from the E.S.E. The land bore N.N.W., distance twenty-five miles.

At four A.M. the ship was kept off N.W. by N. under her three topsails. At daylight the land could be seen about ten or fifteen miles off. The foresail was dropped from its buntlines and clewgarnets (in other sails but the courses, these are called clewlines). No sail was made until the Captain came back on deck in time for breakfast, which is seven A.M.

After breakfast, the Captain gave orders to set all sail and told the Mate to get the anchors off the bows and the chains up. All sail being set and yards trimmed, the Mate, 3d Officer and all the boat-steerers, with the Larboard watch, went forward, as is usual when getting the anchors off the bows. The 2d and 4th Mates, with the Starboard watch, went to work in getting up the chain cables.

In most all whale ships these are run below, in large boxes built of three-inch plank, from the kelson to the under side of the lower deck, one each side of the mainmast. The upper plank on the after side of both is left off, for a man to get into them and stow them when they are run below through cast-iron pipes set in the decks directly over each bin. Where the chains are very heavy, a tackle is put aloft over each hole, a strap clasped on the chain, the lower block hooked on this, a length is hoisted aloft, and a handspike jammed in the pipe hard enough to hold the chain from running back through the hole into the bin again. The word "let go" is given and down comes four or five hundred pounds of chain cable to the deck with a deafening crash. This is pursued until the cable is all on deck. In the meantime, the men have taken it by the end to shackle on the anchor, forward, passing it three times around the barrel of the windlass, out through the hawse pipe, and up to the ring of the anchor, the starboard chain on its side and larboard on the side it belongs. The range of chain, for the depth of water the anchor is dropped in, is laid in lengths close to the windlass. The balance of the chain is arranged in long lengths abreast the main hatchway to the afterpart of the tryworks, neatly and close together.

By the time all was completed, the ship had drawn in toward the land. We hauled on a course directly towards the island until we struck eight fathoms of water, and then let go the starboard anchor, some half-mile from the shore in an open roadstead, but well sheltered from the S.E. trades, which blow from that quarter nine months of the year in these Latitudes.

After paying out about forty-five fathoms of chain, all sail was furled and tackles got aloft to break out the hatchways, fore and main, for empty casks to fill with water, and also to make ready for stowing it away when brought off from shore, as we should need about two hundred barrels. By the time we had taken on board what water was needed and some fifteen or twenty boatloads of wood, and had given the ship a light coat of paint on the outside, some six or eight days had been passed away here, and the men were given two-days' liberty to each watch.

The day before the first watch went on shore for twenty-four hours' liberty, the cask that held the calico that was used by the men to trade with the natives for shells, fruit, and other things was hoisted on deck, and the head from it removed. Each man was allowed so many yards at the rate of twenty-five cents per yard. This he was charged with by the Captain in a book, to be settled for at the end of the voyage. The men could dispose of their cloth in one day or they could make it last for the two-days' liberty, as it would be all the liberty money they could get, with a couple of pounds of tobacco to each one. Tobacco and cloth were about the only trade used at this place.

The officers and boat-steerers could take as much calico and tobacco as they saw fit, at the same terms. To me it seemed funny that these natives on this island, which is no great distance from the Fijis, should be so different. These people have a yellowish-colored skin, straight black hair, and regular features, are peaceful, and in a certain way quite intelligent.

A Catholic priest has been living on the island for a number of years, and has instructed some and made a little reform in their morals, with still a big field left for him to work at during his spare time.

The 4th Mate, with MacCoy (the other boat-steerer in our watch) and myself, on our second liberty day, started for another village along the beach, some three or four miles from where the ship lay. We had heard that the natives in that village had quantities of shells. As there was no road in the jungle back from the beach, the only course we could take to reach it was along the shore, and as the sun poured down hot on the sand, we had to make a break every now and then into the shrubbery, that grew thick a short distance back.

Now and then we would run afoul of a grass house stowed away amongst the trees. Without an exception, a kindly greeting was always extended to us by the occupants. Bananas, young coconuts, pawpaws, guavas, oranges, pineapples and mangoes were offered for our eating. On our leaving we gave them some little presents.

In some of the houses we visited while there, women could be seen employed braiding mats of the finest workmanship in that line that the world can produce. The material used is a peculiar sort of rush, very tough and elastic. This rush can be stripped into fine lengths of an uniform size, not larger than twine; and mats are braided from fifteen to twenty feet square that have the look of neatness that a loom might give. When completed, they can be rolled or folded almost like cloth, without creasing or breaking.

The women, of course, do this work, and none can be seen on this island that have a little finger on either hand, as these are both taken off, at the joint that joins the hand, when they are infants. The reason for it is, that only the three fingers and thumb are used in braiding these mats; and the little fingers would be in the way and entangle amongst the fine material when they were at work.

Just before we arrived at the village we were headed for, we saw a huge pile of stones and lava rocks piled half-masthead high. After stopping a little while to look at it, the other two went on, yelling, "Come on! As if we have not seen piles of lava rock enough since we have been here, without struggling through brush and jungle after more."

After a bit of exertion in making my way through shrubbery and vines, I could see that it was no lava flow that had tumbled this huge pile together, but what use it could be put to I failed to make out. The pile had the shape of a lot of stones tumbled in an immense heap without order or care but to pile up as many that way as possible. After I had taken a look, I came to the conclusion it must mean something; and that if a chance offered, before the ship sailed, I would try and get some information about it.

Struggling back through the brush to the beach, some two hundred feet or more, I could see my companions quite a long stretch ahead, but, as they had taken a number of drinks of awva (sometimes pronounced "kava") they were not making very rapid progress. After walking past a few small houses, we came to one that was of such size it might denote the person living there to be a chief, or some high cockalorum.

This house was some little distance farther back in the bush than most of the others were, standing as it did between large trees whose high limbs joined and formed an arch over it, making a very pretty picture. As if by common consent, we all turned towards it and were duly grateful for the cool shade it offered when we stretched our hot bodies on the verandah, and for the hospitable manner in which we were greeted by the inmates, with offers of fruit, water, and anything they thought we wanted.

The inside of the house was scrupulously clean. At either end, extending across, it had a framework built some two or three feet above the ground, which was the floor, and nicely covered with mats. These frames answered the purpose of beds. They were about seven feet wide and mats were piled on them at least a foot in depth: at the bottom a very coarse one, the next a bit finer, until the top layer finished with one of those remarkable fine ones I have spoken about. This is the one the parties using the bed sleep on. The weather is so warm here that they need nothing but a postage stamp, if that, for covering.

A string is stretched from side to side at the point where the roof projects over the side, parallel with the edge of the bed, to which is attached a curtain that covers the sleeping apartments, like a drop scene in a show; and a person taking a nap during the day would be free from observation.

The curtains in this house were of the most alarming pattern, of impossible flowers and birds, done in red and blue, that would frighten an artist out of his boots but was well chosen by some whaling captain to fetch more hogs and other recruits per yard than the finest silk would, here. The verandah extended the length of the front and was a continuation of the roof. This was about ten feet wide on the floor, which was the ground, covered five or six inches in depth with fine white shells and sand, with coarse mats for covering. The extreme length of the whole thing was at least forty feet; and as the width of the inside of the house was about sixteen feet, it took a whole haymow to thatch it.

After we had eaten fruit, quenched our thirst with some fine water, and had a rest, the 4th Mate and MacCoy made the old chief understand that they would like some awva. He seemed more than pleased to grant their wish.

During the time we had been here, lots of natives of all sizes and both sexes had flocked in front of the house, most of them sitting on the ground a short distance away, and some of the more favored ones on the verandah. The old chief spoke a few words to a number of the girls between the ages of twelve and sixteen, and some six or eight fine-looking ones advanced from the crowd. Entering on the verandah, they went by us through the doorway into the house. Squatting down on their haunches, they made a seat for themselves on the mats in the body of the house, with their plump legs crossed in front of them—not a very dignified position, to be sure, but a sight which no good sailor would find any fault with. The only dress they had on was a strip of cloth or tapa about a yard wide, wound around their hips with two or three turns, and a string of beads or a wreath of some sweet-smelling green leaves around the neck. Their smooth yellow skin glistened with cleanness. Now and then they would glance at us with a roguish look out of their jet-black eyes.

These girls were all "white heads," so called by the sailors on account of a band across the forward part of the head, composed of arrowroot flour mixed with water, like thick paste. The hair above the forehead was cut short and this mess plastered on, covering the hair thickly from temple to temple, from a line where the hair ended to some three inches back. This was put on fresh every morning as long as they were unmarried, or retained their purity.

By the time the girls were seated in a circle, a large

wooden bowl had been placed in the center of the group and a man was passing around the outside of them with a bunch of knotty roots and leaving one or more to each girl as he passed along, until all were supplied. Breaking off a piece from these roots and cramming it into her mouth, each girl set her jaws at work in a lively manner. As one would chew long enough to pulverize it almost as fine as a wad of okum, she would drop it out of her mouth into her hand, slap it into the bowl, and reach out and break another piece from her root. This factory kept to its work until the bowl was about two-thirds filled with the cuds. Then the bowl was taken by a man to another part of the room, and water poured into it until it would just allow the man to knead the mess thoroughly with his fingers without running over. When thoroughly manipulated, it was strained through some of the fibres of a coconut tree into shells of coconuts, and handed around. About the time the nectar was to be served, I took a walk outside, as I had no use for any of that as a beverage.

After visiting a number of houses and buying as many shells as I could carry, I started back to where I had left my companions. Not long after I started to retrace my footsteps, a native came up to me and pointed towards a house some little distance back from the beach, in an opening amongst the trees, saying rapidly something in his language that I could not understand; but by his signs I made out that he wanted me to go to the house he was pointing at. As I had all the shells I could well carry, I made him understand that I was not wanting any more. He shook his head, as much as to say, "Not that," but kept urging me to follow him. Thinking there might be some entertainment got out of it, I followed him, he offering to carry my shells, which, with the generous spirit that is so largely developed in me, I kindly allowed him to do.

Striking into a well-worn path that wound in and out by

patches of awva, under trees that sent out wide-spreading branches that completely hid the sun's rays with their leaves, we neared the house that stood on rising ground some half-mile back from the beach.

I wondered in my mind what it all meant. If it was some of the Spanish places I had been in, I would have made up my mind this might be an intrigue or a knife, and perhaps both; but knives were not common here and the other was not necessary. Still, I did not know but what some young daughter of a high chief, who wanted a bold American for a husband, might be prepared to make me an offer; but this flattering idea was quickly dispelled as we approached the house and I was met at the doorway by an old Catholic priest, who bid me welcome in broken English, and invited me to enter.

A rough, small, square table stood in the center of the room, with two seats made from pieces of pine boards that had been obtained from some whale ship calling here. A bookcase made from a packing case stood against the side of the house, covered in front with a piece of cheap calico, over which was a shelf fastened to the wall. On this rested a small boxlike affair in which stood a gaudy painted figure of the Virgin Mary, with a face and eyes so mournful that one might suppose she was just ready to cry. On the walls, here and there, were highly colored pictures of bleeding hearts with an arrow piercing them, and Christ on the cross, with numbers of others of the like kind.

I took one of the seats he offered with the politeness of a French gentleman, and he informed me in broken English, which, for fear I might render this writing entirely obscure, I will not try to imitate, that he was in need of some medicine. Some of the natives had told him that I was buying shells in the village, and he had sent for me to see if he could get me to help him by sending a native to the ship. (This he did the next day, and the Captain gave him quite a number of different kinds of medicine, after my telling him about the priest's trouble.)

After making known to me his wants and offering apologies too numerous to mention, he brought out from somewhere in back a glass jar full of a bright red-colored liquid. Setting this on the table, he removed a large stopple from the jar and poured out a couple of glasses. He offered me one, telling me that it was made from the juice of prickly pears. "Although it has not the fine flavoor ob de wines ob la belle France," he said, it was the best he had to offer me. After bowing to each other we placed the glasses to our lips. I found the decoction not so bad but what it tasted better than the tepid water one had to drink here, and the beautiful color added much to the taste.

After we had imbibed two or three times, I told him that I must find my companions and we must work back to where the boat landed from the ship, as it would be sunset by the time we arrived there. As I started to leave, he wanted me to join him in one more glass, but I did not know how this stuff might affect my steering qualities, so I told him that I thought I had done well and had enough.

"Dat is all right," he replied, "but de toaste I veree muche wante you and me to drink for de last one time, we nevar more see one anoder."

Filling our glasses again, he gave his toast: "America and France, the two greatest nations of the earth!" Bowing in due acknowledgment to this, I finished my glass, and following his example, I turned my tumbler bottomside up on the table. No doubt it would have been in the due order of such things for me to have proposed a toast in return; but if that thing started, the Lord only knows when I would have reached the ship, as I could feel the stuff going a little to my head.

He came with me a short distance on the path that led towards the beach, with the native, carrying my mat basket of shells, walking behind us. Just before we separated, I happened to see a pile of lava rocks, and this put me in mind of the huge pile I had seen and wanted information about. Thinking the priest might be able to enlighten me on the matter, I explained the wish to him. He laughed most heartily for a minute or two, and then motioned me to a seat on a flat piece of lava rock.

Some few years back, so he told me, a chief lived near where that pile of rocks stands. He was very bad. He would kill natives, beat them, take their wives and any of their daughters he might fancy, and do everything that was bad. One time he was taken sick. After a time it was reported he was dead. The natives flocked to the house in numbers. Men, women and children held high jinks over the joyful news, but right in the height of the rejoicing he came crawling out from under the mats thrown over him. It seemed he had some disease that had rendered him only insensible for a short time.

After getting well and around again, he played the mischief worse than ever, and if bad before this, he was much worse after finding out how pleased everybody had been at his supposed death. This went on for a time. When he had got the people to such a stage that they did not know what to do, he was taken sick again. From this he never recovered, but died sure enough. At the time of his death no one said a word, but all kept quiet and let him lie until such time as he was commencing to decompose. Then, approaching where he lay, very cautiously, some men, with a rope that led quite a distance outside the house, made one end fast to his legs. When given word, those on the outside pulled on it, as they ran, until they reached a hole that had been secretly dug; and they tumbled the body head over heels into it. A crowd of people concealed in the bushes then suddenly appeared and with haste piled stones into the grave as fast as they could, until a ton or more was over him. Not being satisfied with that, and having some fear he might get out, they more leisurely piled the mound or hill over where he lies.

Thanking him for his information, I shook hands, and taking my shells from the native, we parted. Steering as straight a course as possible under the heavy press of the priest's hospitality, I came to the house where I had left my shipmates. As I came into the yard in front of the house, but few natives could be seen anywhere around. Everything was very quiet. I went into the house, and there a sight met my gaze that explained itself. Too much awva had overpowered chief, sailors, and some favored few of both sexes. The two platforms answering the purpose of beds were well occupied. The female portion made quite a display, as their drapery at the best was not extensive.

Hunting over the ruins for the 4th Mate and MacCoy, I succeeded in rousing them out of the stupid sleep that the awva had produced, and made them understand, after much effort, that it was time for us to be starting for the boat, unless they intended to pass the night there. We then started along the beach towards the landing. Each of them had a mat basket with a lot of shells in, but very few that were worth the trouble of packing. These they had bought when under the influence of the awva, from natives who were sharp enough to know they could not tell a good shell from a bad one.

I had found a short stick; and putting one end of it through the beckets on two sides of my basket and the stick over my shoulder, I found it much easier to carry that way. Letting the others go on ahead a short distance, I followed after, and had more fun than would fill a scrap tub twice over, watching them stumbling along with their shells swinging fore and aft. Briggs, the 4th Mate, was now and then making some heavy-weather rolls and lee lurches; and after we had traveled a mile or more, I noticed that the bottom of his basket was opening.

Keeping Briggs ahead, I could see the basket open along the bottom more and more. Now and then a shell would drop out, until after a bit the thing opened enough to discharge them suddenly and throw him out of balance. After tangling his legs once or twice, he managed to sit down quite gracefully on his left shoulder and roll completely over on his back, saying, "There now! see that!" with a tone of voice that might imply he would like to see if either one of us could do it as well as he had.

After a few minutes he seemed not to want to get up but was inclined to go to sleep. Rousing him, after his swearing that he would see me "in the shades of some place more warm," we forged ahead, I keeping in front now to try and hurry them along. We went on very well until, seeing a house in the bushes not far from the beach, nothing would answer for them but to try for another drink of awva. They started for the house and I followed them, trying all the time to induce them to come on towards the landing. But it was no use; awva was what they wanted, and awva was what they were bound to have.

The house was a small-sized one, and on our entering it, no one was in it or around, as far as we could see, except an old woman. She was sitting on some mats in the center of the house, with lots of rushes on the floor around her, which she was stripping into shreds for braiding into mats. This old woman had seen many a year roll by and helped grind up the roots of the delectable awva. But little of the grinding tools were left to do much in that line, as could be seen when she opened her mouth and pointed her finger towards the cavity, which looked like a burnt-out cellar. Only a few yellow-looking fangs could be seen here and there in it. At the same time she shook her head, as much as to say, "You see how I am fixed," when those fellows made her understand that they wanted her to chew them some awva.

They made many signs and offers of fishhooks, tobacco and a couple of gaudy handkerchiefs that they had left; and finally the old hag, who had two wrinkled-up dairy farms that hung so far down in front that she could have thrown either one over her shoulder, started to chew up the horrid stuff. Much disgusted with them, I left, telling them that there might be worse fools but I did not believe it possible.

They appeared at sunrise next morning, on the beach where the boat landed to take off those who had passed the night on shore, looking much the worse for wear. The liberty being ended, the anchor was hove up and we stood to sea, bound on a short cruise to the Line.

There grows on that island a species of curry root which adds a fine flavor to stews and soups, when cut into small pieces the way carrots and turnips are. This also gives a bright rich-yellow color to the mess. The natives also use the root by pounding it up and mixing some water with it, and rubbing this over their naked bodies from head to heels; and when well smeared with it (as yellow is their favorite color) they feel as proud of their appearance as a dog with two tails, and seem happy when rubbing some of it off their bodies onto whatever they come in contact with. A person can hardly touch anything but what he will show the marks. Some of our men showed yellow on their skins and clothes for a week or two after we left there.

Stove In

HEN in about the Lat. of 5° South, Long. 170° East, about a month after we left Rotumah, all hands were made happy one morning, just as seven bells struck, by the pleasing sound from the masthead of "T-h-e-r-e s-h-e b-l-o-w-s!" The answer to the Captain's question of "Where away? And how far off?" being "Four points on the lee bow, two and a half miles off," the mainyard was hove aback, lines were placed in the boats, and all hands but those to keep ship went to a hurried breakfast, which was soon finished, and the boats lowered away.

The weather for the past two or three days had been squally. Squalls of wind and rain would blow for an hour at a time, with rain coming down in torrents, obscuring everything but a narrow circle of darkened waters half a mile around the ship. When the squall would pass over, the sun would shine out and the sky become quite clear until another squall forming would cover all up again, sometimes two or three hours apart, at other times not so long. It was in between two of these squalls that the whale we lowered for was sighted; and the weather continued so good that before another squall came on we caught sight of him again: a large lone bull whale, who no doubt had been driven off from a school of cows by one more vigorous than he.

Two large bull whales in one school of whales at the same time is unknown without they would be fighting, when the conquered one would leave instantly. They fight with terrific fury, using their jaws in rushing at each other, something like two men fencing with swords; and cases have been known of whales having their jaws twisted quite out of shape and sometimes broken entirely asunder. I have seen one or two instances of the first kind myself.

Before the whale went down again, we had got so good a run of him that by the time the next squall cleared off we sighted him no great distance from our boat, and although drenched to the skin by the rain, we started for him with glee. As we were some two or three ship's lengths nearer to him when he broke water than any other boat, and square in after him, the chance was ours, and with our sail and oars we improved it for all it was worth.

The whale was heading to the leeward with the wind about four points on his quarter, and as there was a fresh breeze blowing, we shot up across the corner of his flukes with a free sheet, and when the boat's stern passed clear of his flukes a quick stroke with the steering oar caused the boat's head to shoot to windward (as it is always the rule to go on the lee side of a sperm whale when approaching to strike or lance). Bringing the boat's head within two or three feet of his side, I let him have the first iron, which I had been holding in my hand since ordered to stand up by the 2d Mate when crossing the whale's flukes. The second iron was planted into his huge body, no great distance from the first, a second after. Throwing overboard the coil of stray line, I turned around and commenced to roll up the sail.

This job was none too easy, as the whale, instead of sounding as is almost always the first thing they do when fastened to, took to running to the windward. This caused the sail to slat and bang so much that at times the sprit would slip out of my hands when I had the folds of the sail wound around it almost to the mast, and away the whole thing would fly, adrift again. With the assistance of the bow oarsman, I secured it at last; and unshipping the mast out of the thwart, I passed it to the 2d Mate, who shoved the heel of the mast under the after thwart, with the upper part resting on the gunwale and the covering board over the stern sheets on the starboard side of the boat, all clear of the line and steering oar, leaving everything clear for working on the whale.

As soon as the mast was out of the way, the 2d Mate took his place in the head of the boat and I went aft to the steering oar. By the time everything was straightened in the boat, we had run quite a distance away from the other boats and were as far to the windward as the ship, but some mile and a half ahead of her, the whale showing no signs of slacking his speed; so all we could do was to hold our line, trim boat and keep the water bailed out that now and then rolled over her sides and bows when she pitched into a larger sea than common.

This thing went on for an hour or more, when big dark masses of clouds could be seen to the windward, making up into a mess that denoted a squall of unusual severity and continuance. The ship saw it and took in her main-topgallant sail and flying jib.

Not much had been said out of the usual line by either the 2d Mate or myself up to this time, but now we both expressed ourselves in regard to the whale and the look of the weather. If the confounded whale would only give us a show to lance him, though, we did not care much about what might be in that impenetrable mass of black matter that was fast rolling down on our unprotected heads from the windward.

About the time the clouds had gathered overhead so thick that all signs of blue had vanished and the first pattering drops of rain began to hit us, the whale milled a little to the leeward; and by the time full force of the squall was on us, he was running at the rate of ten knots an hour with the wind abaft the starboard beam. Keeping this course and speed for half an hour, he suddenly stopped and commenced to roll and tumble, snap his jaws and strike with his flukes. This was what we wanted, in one way, since now might be a chance to get a lance into him; whereas there would be none when he was running like he had been.

But of all the different views of whaling that it has been my lot to see, this beat them all. Here we were in the midst of a raging tempest, rain pouring down in torrents, the sea and wind combined in a roar that was accompanied now and then by the sound of the whale's flukes as he struck the water, after raising them fifteen or twenty feet above the surface, sending the foam flying in every direction when he brought them down in his rage.

All the world was lost to our view, a little space in the world of waters in which a battle was going on; a whaleboat with six men in it, a whale and a gale, each trying to win. We did not pay much heed to anything but the whale, and watching a chance the whale gave us, we sent the boat ahead and with a skilled aim the 2d Mate drove his lance deep into the life of the whale, and the thick blood that came pouring out from his spout hole told of our battle won. "Stern all! Stern all hard!!" was the cry from both of us to the men, as the whale turned towards us with his head and just missed the boat with his wide-open jaw. Sterning off a safe distance from him, as he was tumbling about too much for us to do any fooling just then, we watched him in his struggles for a few minutes. Raising his head at last well out of water, he shot ahead and turned flukes.

"Look out for the line!" was the order from the 2d Mate to me. "He is going to sound."

"Aye, aye, sir," was my reply and at the same time I took an extra turn around the loggerhead and reached for a canvas nipper to shield my hands from burning by the friction of the line passing through them when grasping it to prevent its running any faster around the loggerhead in the stern of the boat than safety obliged.

Down he went, dragging flake after flake out of the tub until more than one-half had rapidly disappeared; I holding the line at times hard enough to pitch the head of the boat under water, and getting a sharp word of warning to be careful from the officer. Still the line ran out, and the 2d Mate was calling for the drags. (These are boards nailed together, about eighteen inches square, with a block of wood in the center, through which a hole is bored for splicing a piece of towline long enough to fasten around the whale line with a rolling hitch. They answer to retard the whale's progress, in a degree, when there is danger of his taking a boat's line.) The 2d Mate fastened on the two drags we had in the boat. By the time he had done this, our line was run out to the last two flakes; about one hundred and eighty fathoms being gone of the two hundred and twenty that is the length of a whale line.

We commenced to look blue, as, if he took our line, we should have but little chance of ever seeing any more of him. We knew he had his death wound; and if he did not take our line, when he came again to the surface his life would be short. Just then the strain on the line slackened. This made our hearts rejoice, and when our officer gave the order "Haul line," every man faced around forward on his thwart, and with both hands pulled hard together and brought a little line back into the boat, gaining on it faster and faster as the whale approached the surface, which he now was doing. Coming to the drags, the lanyards to them were cast off from where they had been made fast to the line, and they were taken into the boat.

By the time he broke water, some three or four hundred feet from us, we had hauled into the boat about two-thirds of our line, and as he was only moving very slowly through the water the boys hauled so lively on the line that I had hard work to coil it fast enough in the open part of the stern sheets to keep up with them. The whale was throwing out thick clots of blood by the barrelful at each spouting, and hardly moving through the water, and the boat shot towards him as fast as four men could haul the line.

When we had approached within a short distance of him, the 2d Mate, much to my astonishment, ordered the men to take the oars, and at the same time he told me: "Put the boat on the whale, so that I can lance him again." He and I had had so many bouts with each other that I hardly liked to say anything against any more orders he might give, but in this case I could not keep still, for our lives were to be placed in jeopardy very foolishly.

The look of astonishment he had on his face when he turned from looking forward, towards me at the steering oar, on my saying to him that I thought he had made a mistake, as he certainly could not think under the existing conditions of wanting to lance that whale, would have made me laugh if it were not a too serious time for levity. The men had stopped pulling, no doubt at my great breach of discipline in asking him such a question, so the boat almost stopped.

"Pull ahead!" he yelled at the top of his voice.

"Hold on a minute, Mr. Griffin, let me say a few words first," was my reply, and then as rapidly as possible I called his attention to the fact that the whale was about dead and would soon go into his flurry. Also, that though the worst part of the squall was over, nothing could be seen yet any great distance from the boat and the ship could not tell our position, as the whale had changed his course so much, from what it was when we were last seen by her, that if our boat should get stoven the chance of her finding the whale or us would be mighty slim when the weather cleared up again. "This is all I have to say in the matter, except, if you are bound to lance that whale, let me put the boat on his lee side, instead of going to him as we are now, heading on to windward of him."

The only answer he made to any of the reasons or requests made by me was that he did not think I was such a coward, and to put the boat on to the whale the way she was headed; the quicker I did it, the better.

Telling the men to pull ahead, I told him he would not find me far behind him in any danger that any reasonable man had to face. A few strokes of the oars, and the heave of the seas sent the boat on to the whale. He set his lance over its shoulder blade and, holding by the pole, thrust it up and down two or three times, and then sung out: "Stern all!"

The men did their level best to obey the order, but even with the 2d Mate assisting at the harpooner's oar when he saw the four men failed to stern off the whale, nothing more could be done than to get the boat a short distance away from him to the windward, and then be swept back by the next heavy sea. This happened three or four times, and the boat had worked aft on the whale quite a piece, when the whale, with a blow from his flukes, knocked the bottom half-out of her and tumbled us all into the water.

The 2d Mate, who could not swim a stroke, managed to be amongst the first to grasp hold of some part of the boat and so keep afloat; and before he had hardly got the salt water out of his mouth, he was giving orders for securing the oars and lashing them across the gunwales of the boat, to keep her from rolling over and over in the seaway, and drowning us all. He was a brick, in many ways, when a pinch came.

The situation we were now in was rather critical, to say the least, and how it would end would be hard to say. We had lashed the oars across the boat by pieces of small throat-seizing stuff fastened under each gunwale for just such an emergency as this, and so we kept the boat on her keel, thus affording us a good resting-place for our arms over the gunwales on each side. Of course, our bodies and legs hung in the water, which would afford an easy meal for the sharks that abound in the sea in those Latitudes. Of these we felt somewhat nervous, but had hopes the whale would attract them by the blood from his carcass, that lay dead on the water some hundred yards away; for soon after he stove us, he had gone into his flurry and turned fin out.

Just before we took our bath, the clouds began to break, and an hour afterwards the sun was shining brightly. The wind went down to a light royal breeze and the sea in two hours was so smooth that it ceased to break over us. The last squall had been the most severe of any in the last twenty-four hours and seemed to have exhausted itself and brought good weather with it, much to our satisfaction in more ways than one. When the weather cleared, the ship could be seen some eight or ten miles away, but no signs of any boats to cheer us with hopes of relief. They had lost all run of us as well as the ship had, and when it cleared up they went to the ship and added their eyes to the others on board, in the vain effort to catch sight of our boat on the waste of waters.

As the time we got stove was about ten A.M., we felt not much apprehension but that long before dark we would be found. We made the best of our surroundings. The water was not very cold, and until our legs became almost as wrinkled as a washerwoman's thumb, we suffered but little that way; but the sun shone hot on our heads and we suffered with thirst. Now and then the sharp back fin of a shark could be seen cutting through the water no great ways off, but as none took a bite at us, we lost, to a certain extent, our fear of them, after a time.

But the few hours of hope and despair that I and the rest passed through was enough to answer a lifetime. Many were the regrets the 2d Mate expressed that he had not heeded my warning, and let the whale alone. Telling him that it was all right, it might have happened anyway, seemed to afford no relief to his feelings.

When we first saw the ship after the squall was over, she was on the starboard tack, heading towards us. If she kept on that course, we thought she would pass about a mile or two by us to the windward, as we were about two points on her lee bow, and some one of the many eyes on board her might see the waifs we had stuck up, one in each end of the boat; for it would be impossible for the boat, level with the water, to be seen at the distance she would pass.

With eager and longing eyes and hopeful hearts we saw her approach close-hauled on the wind, with all sail set. On, on she came, her hull fast raising above the horizon, then her white waist showing more plainly, until the upper course of her bright copper could be seen under the lee bow as she rose on a sea and, settling down, sent a sheet of foam dashing away ahead, under the dolphin striker. "She will soon see us now," we said to each other, as men could be seen on the lookout from the royal yards, topgallant and topsail yards also. The time had now arrived when every minute we expected to see her swing her head off from the wind and point for us. Our hopes were doomed to disappointment, for just then we saw the jib sheets ease off, and when they commenced to slat, we knew she was going in stays and soon would be on the other tack and standing away from us. No one but him who has been clinging to some frail support on the wide ocean can tell of the agony of such moments.

When her headyards were braced full, and fore and main

tacks boarded on the tack, so that every foot she went through the water was directly away from us, each man gazed at one another but no words were said. Every man had the same thoughts of how little chance for us was left, as in tacking ship those on board of her believed we must be somewhere to the windward; for we had been seen going that way before the squall struck. But here we were to the leeward; and they, not thinking so, would not keep so bright an outlook that way, naturally, as to the windward; and the chances were she never would come so near us again.

With hearts that seemed like they would burst, we watched her sail away, so stately and silent, until her hull had settled below the line of the horizon. But then we were gladdened to see her tack again; and although we knew that if she hugged the wind on that tack, she would pass by us some miles to the windward, still, this would be better than having her stern towards us. Besides, it showed that our Captain knew he had run off the line he ought to find us in, and that he would look to leeward, after beating across it to the windward, far enough to satisfy himself we were not in that direction.

They passed by again and stood quite a distance on before she went about. When passing us on the port tack this time, only the head of her topsails could be seen; and by the time she got ready to tack again, which she did, she could hardly be seen by us and about all our hopes of rescue were gone. Still we strained our eyes, that had now become quite painful with the spray flying into them and the heat of the sun, to catch the gleam of her white sails, so far away to the windward. I asked the 2d Mate if he could tell the time by the sun. He in a choked voice said he thought it was about three P.M. If so, we must have been five hours hanging on this boat and we only had about three hours more of daylight, as the sun would set at six. If we were not picked up before that time, how few of us would ever see the sun rise again?

These thoughts passing through my mind, with other thoughts of home, and how bad my poor mother and sister would feel when the loss of this boat's crew was known, caused me to lose myself partly to the surroundings that existed; and I was suddenly aroused by an exclamation from the blacksmith, who pulled our bow oar, that the ship was running off nearly before the wind.

Looking towards her, it could be seen she had come in sight more plainly, and in a short time there could be no doubt but what she was running off with the wind on her quarter and heading on a course that would carry her no great distance on the other side of us.

Up went our water-soaked spirits and never were poor devils more pleased than us, to see the rise of her full-bosomed sails, foot by foot, as she came rolling and heading nearly towards us. When within three or four miles of us, for some cause unaccounted for, she luffed two points and this carried her so far by us, to the windward, that she was farther off when she passed than she was the other time she came near. Down went our thermometers, and cold chills of despair went through our bodies which the long soaking in the water so intensified that we felt like giving up.

During the last two or three hours of our misery, not much had been said between any of us, and so indifferent had we become to our condition that a shark's fin would have to appear pretty close to us to cause much excitement or any splashing in the water with our feet to frighten them off. So, when one of the sharks caught the blade of our steering oar (which was hanging in the becket at the stern, by a pin in the handle, and the blade some twenty feet away) in his teeth and gave it a heavy shake, at the same time splitting a piece off it, so deadened had we become that we hardly noticed the occurrence, or the remark made by one of the men: "Some of our legs will catch it next."

A strange fact in connection with that event is, that at the time it occurred, I had far less horror of being eaten alive by the sharks than I have had many times since, when thinking it over, or starting out of a sound sleep with perspiration streaming, by dreaming of the cold-gray look the round devilish eyes of some of those sharks had. If one of us had been bitten, short work would have been made of the rest.

The ship kept on her course away from us for fifteen or twenty miles longer. Then we saw her keep off before the wind. This course took her no farther from us, which encouraged us; but if she kept it up it would bring us to the windward, and if she hauled on the starboard tack, all hope of their finding us before morning, if then, would be gone.

I had been thinking for some time what we could do to attract her attention, and it now struck me she was in such a position that if our sail could be raised without the boat rolling over, the width of it could be seen a long way off, and, as we were off her beam, some eyes would catch sight of it before it tumbled down or rolled the boat over. I told my thoughts to the 2d Mate, who said the idea was good, if we could only get the sail up. The trouble of that was, the boat would roll over, bottomside up, if great care was not used; and if she did, some of us might lose the number of our mess. But we would try it anyway, he said, as it might as well be ended that way as any other.

By some exertion the sail was worked out from under the after thwart where the lower part of the mast had been shoved, and was lifted on top of the oars that were lashed from side to side, the heel of the mast being over the hole in the bow thwart. Cutting a short piece of lance rope some two fathoms long, this was fastened with a clove hitch, about two feet from the end of the mast that fitted into the step on the boat's keel. The two ends of this heel rope were made fast to the midship thwart, one on each side, as near as possible where the ends of the thwart fastened to the boat's side. The sail was unrolled. Another piece of lance warp was fastened to the top of the mast, with ends long enough to reach the sides when the mast would be upright, with one man on each side to tend these; this to answer as backstays, to keep the mast from tumbling either side. The rope that fastened the sail when rolled up was led forward, and the end of it rove through the eyebolt in the head of the boat that the hook of the tackle to hoist her with fitted into. Taking a short piece of rope with me to the head of the boat, I made with it a lashing for myself to sit in; and when in position, I had one leg each side of the boat's bows and one arm free to work with.

The end of the rope passed through the eye bolt was taken hold of by a man hanging onto the harpooner's oar, and a round turn was taken with it to hold what might be gained as the mast arose. The end of the sprit pole was cut from the corner of the sail and fastened to the end of the mast; this to help lift the mast upright, as it would not do to have a man in the boat to help.

The 2d Mate and the after oarsman, holding each with one hand on the boat and using the other to shove the pole with, got as far aft on the boat as they could work, and the word was given to lift. After trying three or four times, we made out to raise the sail two-thirds upright, and although it seemed the boat must roll over, she did not. One thing in our favor was that she rode head to the sea.

During the time we were working to raise the sail, little attention had been paid to the ship. But when the sail was blowing out and showed a signal of white drilling at least fifteen by six feet, that ought to be seen plainly on the ocean eight or ten miles on a day like this, we all turned our eyes in her direction. The sail fluttered in the light gale, and the boat rolled enough at times to frighten us that our signal might tumble over before it was seen. I do not suppose ten minutes had elapsed from the time we had let our banner fly to the winds, but minutes seemed like hours; and it seemed that long, before down went her wheel and round came her noble head, with her jibboom heading straight to us.

We yelled, we shouted and laughed. Cold and sharks were nothing now to us. With blue lips and drawn faces some of the boys tried to make a joke, but it ended in a sob that was almost a fresh-water cry. When all doubt of their losing run of us had passed, we let the mast down for fear the boat might roll over and cause a tragedy.

The old ship came bowling along. Oh, how good she looked! To me she seemed the biggest spot on earth or ocean, and to once more tread her white decks would seem bliss indeed. Soon the white foam under her forefoot could be seen rolling and tumbling as she dashed her sharp bows into the seas that attempted to stop her on her mission of mercy; she in scorn smashing them into bubbles and suds that went dancing along her sides into the wake astern, with hissing sounds that might mean regret at their impotency.

When some half-mile from us, she hauled up her mainsail and laid the main-topsail to the mast, and lowered three boats, one of which went to the whale. (It was about a half-mile from us, we had drifted apart.) The other two boats came to us. One took us into it and the other commenced to clear the stoven boat of the oars and sail, to take it alongside the ship. The ship ran towards us as soon as they saw us leave the stoven boat, so we had but a short pull. When we arrived alongside, most of the boys had to be hauled on deck with a rope. Hot coffee was given us and we were sent below to change clothes, after which we turned in and had a rest for an hour or two, to set our blood in motion.

While we were turned in, the stoven boat was taken aboard and the whale hauled alongside, sail taken in, and all made snug for the night, as by this time it was dark. After we had our suppers, and were smoking our pipes, leaning over the bulwarks with our elbows on the rail, looking at the whale alongside, the other boat-steerers allowed that they had given us up, and had no hopes of finding us. The Captain thought we must have been taken down by the whale; and when running off before the wind, just before they caught sight of our half-set boat sail, he had said he did not know which way to look for us now, or which tack to haul on the wind, as the ship was some way to the leeward of where the boat had last been seen. "So you see, old fellow," said one of them, "if we had not caught sight of you when we did and had run a short distance more to the leeward, we would have missed you on either tack we hauled to the wind on, and the rest of us fellows would have had a chance to have bought some of your old clothes and your chest, cheap."

"So you might," was my reply.

It matters not, to a thorough English or Yankee sailor, how serious or startling a thing might be. If passed safely through, they will have some joke about it.

It took me some days to recover from the stiffness in my bones and muscles, but I was never laid up for it. The next morning at daylight, the tackles were put over the side and we commenced to cut. As the whale was large, it took us until noon to finish cutting him in and bailing the case. After dinner the tryworks were started and the stoven boat placed overhead on the skids, bottom up, and another put on our cranes until ours was mended. Her keel was not broken but the bottom on both sides was badly stove in, and the carpenter had to use lots of cedar boards before he made her as good as she was before. It took us forty-eight hours to try out the whale. He turned up in the casks oil enough to make one hundred and ten barrels.

AFTER stowing down the oil from the last whale, we took a short cruise through the Kingsmill Group over some of the same waters that we had been through before, but without seeing a whale; and then we hauled the ship to the South for New Zealand, there to take our last cruise and then start for home.

How sweet that name sounds, to one who has cruised for weary months on a sperm-whale ship, none can tell but him who has been perched aloft over the main-royal yard for weeks, gazing to catch sight of a breach or spout, some days in calms, some days in rain, and others swaying back and forth between heaven above and the deck, a hundred or more feet below, as the old ship plunges, at times, under double-reefed topsails, into a head-beat sea that sends the spray flying from her weather bow, while the wind catching it sends it back over the rail, much to the disgust of the men who are in its path.

About the first of January we arrived back to the old cruising ground off the Three Kings. For the most part of the month, we had one gale after another that made it very hard for us to see a whale or to have any comfort. We sighted whales two or three times, struck one and lost a line, the weather being so bad that he, running to the windward, would have swamped the boat if the line had not been let run. Striking one more, the irons drew from him; and so ended our last whaling in the Pacific.

On the 25th of January, 1853, we let go our anchor in the arm of the Bay of Islands, where the English garrison is stationed, called Waapoo (as we sailors pronounce it). This place is some five or six miles from the town of Russell, the place where we anchored before. The Captain, I think, chose this place on account of feeling downhearted at our hard luck, as quite a number of ships that had done well were lying at anchor at Russell.

Amongst the ships lying to anchor off the town of Russell, was one by the name of *Brighton,* a Northwestman; as the name signifies, a right-whale ship, cruising on the

Northwest coast of America. This ship had taken quite a quantity of oil, and as he intended taking another cruise North, he made arrangements with our ship to take about one thousand barrels of his oil, on freight, to New Bedford. Ships often send oil home, when a chance offers, as it gives them more room and places the oil on the market sometimes two or three years before the ship that sent it arrives back to the port she sailed from.

When we were ready to take her oil, our Captain informed the master of the *Brighton*. He got his ship under way, and sailed up the bay and anchored near us. We were about a week taking her oil on board and stowing it. After this was finished, we took on board about two hundred and fifty barrels of water and some two or three tons of Kaurie gum. The Kaurie gum was for some varnish factory in the States, and, I believe, was the first shipment sent to America; though large shipments had been made to England many times previous.

It is pretty stuff to look at, almost transparent, of a rich amber color, quite light in weight and somewhat brittle, and is found under the surface of the ground. Some of these spots are surrounded by dense forests but others are on open land that shows no sign of a tree ever being anywhere near it; and as far as any information I could get about it, no one could say that the Kaurie gum came from any tree growing there now. The mode of locating this stuff is with a sharp-pointed rod of iron. This being shoved into the ground to a certain depth, it can be determined if the right place is found or not. When it is located, the earth that covers the gum is removed and bodies of it are found, sometimes of many hundredweight, so I have been informed.

After the oil and water were stowed away, we gave the old ship a thorough scrubbing and painting, from her royal trucks to the water's edge, both inside and out; and when done she looked as pretty and neat as a belle in a ball dress.

During the time we were painting ship, a watch was allowed on shore for liberty for a week or more. As there was not so much to attract sailors around the barracks as there would be at the other anchorage, most of the liberty days were spent at Russell. The watch were allowed a boat to go and come in, so long as they caused no trouble, which the many months passed with Old Sampson had taught them to avoid.

Getting acquainted with one of the officers of the company at the barracks, I spent most of my time on shore there, or in rambling around the hills. One day I took some shells on shore and gave them to the officer's wife, who had been very kind to me; and not long afterwards she had a visit from the wife of the Captain of the Regiment. The officer's wife showed her the shells, and she, never having seen such fine ones, was quite excited over them and asked if she thought I would sell her some. On my next visit at her house, she reported the conversation, and I told her I had none to sell; but that if the Captain's wife would accept a few to place in her cabinet to remember a Yankee sailor by, she would be welcome to them. On my next visit on shore I took her a nice selection, asking the officer to hand them to the Captain's wife, with my compliments.

A day or so afterwards, the garrison boat was seen approaching the ship. Coming alongside, an orderly came up the gangway and presented a note to the Mate, who sung out the next minute for me. When I went aft, he handed me the note with the remark, "You seem to have a lady correspondent, among the other things you have picked up here." Telling him that I had not been idle in my rambles on shore, I stepped to one side and opened the note, and found it to be from the Captain's wife. On reading it, I found it a pressing invitation for me to dine with her on that evening, naming the hour.

The orderly had been waiting, and as it said "Answer," I told him to say I would; but I did not like to do it as the clothes I had for my best were not in keeping to sit down at a swell table. Still, the note was so kind and ladylike, and waived all ceremony, that I could hardly do anything but accept. I would have refused if I had had decent paper to write it on, and would not have kept the lobster-back too long waiting.

The officers and boat-steerers were much excited about the whole transaction. I never gave a thing away, but dressed myself when the time came to do so and asked the Mate for a boat to set me on shore, which he kindly placed at my command, with orders to the men to stay on shore waiting for me, if it took all night.

Well, thinks I to myself, when seated in the boat and pulling towards the shore, if this is not the loudest mess I have ever struck. How it would end made me nervous to think of. The idea of a common boat-steerer on a Yankee whaler being invited to dine with the Captain (and a lord at that, so I had heard he was) of a crack regiment, was a little too-too.

When the boat landed, I gave the men a dollar to get something to drink at the public house, and told them I would call for them there when I was ready for them to take me on board. Then I started towards a neat cottage on a slight hill, where the Captain resided, and coming to a gate, I opened it and walked up a graveled way that led to the house. Trees, shrubs, and flowers lined this the entire way to a broad stairway that led to a wide verandah that ran the whole width of the house. When I reached the foot of the steps, a soldier on each side dropped a gun with fixed bayonet across my further advance and demanded my business.

Just then the front door opened and out came a fine-looking man, except for a look of slight dissipation on his face. He spoke a word to the two soldiers that barred my way, they raised their guns, and I mounted the steps. He shook my hand and told me to walk in. On my entering the house a beautiful lady, whose sweet face was a pleasant spot in my memory for many months after, most kindly bade me welcome, and with many thanks for the fine collection of shells I had sent her, assured me that when they were placed in her home in Old England, she would never look at them but her thoughts would be of the Yankee whaleman who gave them.

In fifteen minutes after taking my seat in that house, I felt perfectly at ease. The true good breeding of those two people swept aside my embarrassment and lack of refinement. Shortly after I arrived, dinner was served and it was the most swell affair that I ever had been guilty of attending. There were no others at the table but us three.

For an hour after we had finished dinner, we had a pleasant conversation, and when they found out "how English" I was, they seemed still more pleased; and the Captain would not consent to my leaving until long after his lady had retired. We smoked, talked and sampled his Scotch whiskey until midnight, and when I told him that I must go, he kindly said he would let me "show one of his orderlies the way" to the boat. There I found the men, as the public house had closed up.

When we got alongside I found the steps of the gangway all right, crawled up on deck, and soon went below and turned in. The effects of that dinner, and the trouble of showing that orderly the way to the boat, did not last over two or three days.

This Captain (whose wife was a rich titled lady) had been by the influence of his friends sent to this far-off station, in hopes that away from London life he would change his habits, as his life there was getting a little too fast. He was said to be a brave officer and complete soldier.

Home around the Horn

THE MORNING of February the 8th, 1853, was bright and pleasant, as were our hearts, when the call was for "All hands, to get the ship under way for home."

The Mate had hardly given the order to man the windlass before the heavy iron brakes were in place, manned by a dozen stout men on either side, eager for the word, "Heave away!" as race horses are for the word "Go."

Soon the clink of the iron palls could be heard as the windlass spun around from the vigorous strokes of the men, who were raising their bodies and arms on one side to their fullest extent, while the opposite side bent nearly to the deck as the handle through the iron beam of the brakes went up and down. The echoes from this and the songs of "High, Randy, O," "Mobile Bay," and "Off She Goes" made the hills around us seem alive with voices and strange noises, and caused many a door and window to open on the shore, no great distance away.

Soon the cry, "Avast heaving," came from the Mate, who was standing on the breast hook, between the knightheads, watching to see when the chain was up-and-down, which would show the anchor apeak and call for sail to be made on the ship.

"Lay aloft, and loose all sail, Starboard watch aft, Larboard watch forward," came the order. Each watch to their stations, men from each spring like monkeys into the riggin', making the shrouds vibrate as their nimble feet fly from rattlin, to rattlin, out on the yards, off with yardarm gaskets, then bunt gaskets, down into the tops and crosstrees, to overhaul the buntlines, reeftackles, and clewlines as the sheets hauled home. At the same time the men on deck at the halyards are hoisting the yards to the masthead. Soon all sail but the jibs are set taut.

"Haul in the port mainbraces and starboard forebraces" is the next order. We want to cast the ship's head to port, as the passage out of this bay is on our larboard bow. "Man the windlass," is the next order now, as there is a breeze blowing against the sails. The chain has a strain on it so that the heaving is hard, but with sturdy arms and strong backs the men wind the chain link by link through the hawse pipe, until the heavy mass of iron breaks its hold in the mud and the ship's head slowly turns to the left.

"Hoist away the jib!" yells the Mate, at the same time telling the men to heave on the windlass lively. As we have a little sea room, the headyards are not swung full until the anchor stock appears in sight above water. Hooking the cat block on the ring of the anchor, fifteen or twenty men soon have it at the cathead. By this time all sail is set, alow and aloft, and with yards checked in we go dashing past the headland on our right. With cheers from the other ships and shore, which we duly answer, we turn the old ship's head towards that spot, so dear to all, that we left forty-four months ago.

The anchors are soon fast on the bows and the chains stowed below. By three P.M. the outline mountains of New Zealand show a dim blue trace astern, soon lost to view.

The ship is ploughing a roaring, tumbling mass of white-foaming water from her bows as she makes at least ten knots an hour with the wind on her starboard quarter, with every sail swelling and straining from the yards as if they would do their best to shorten the distance home.

The wind hauled more aft during the night and freshened, so by daylight the ship was under whole topsails and foresail, with the jibs and mainsail furled, as they could do no good, the wind being too far aft. When the sun came up, it had a cold silvery look as it showed now and then through the openings of the dense masses of steel-colored clouds that denoted by their tumbled confusion a forerunner of a westerly gale. As this would be fair wind for us, none cared, if we could run the ship through it and not have to heave her to.

By dark, though, the gale had increased to such an extent that the mizzin topsail was furled, the fore and main topsails doublereefed, and the ship was dashing madly down the sides of the seas that no other ocean in the known world but this can produce, rolling her rails under water in the trough, her decks flooded with water fore and aft, struggling to free herself from the tons of water on her decks, settling her stern down with bows well raised, masts and yards creaking, sails straining as she rushed to climb the twenty or forty feet of the wall of water ahead.

About six bells the foretack parted; the noise of the link when it gave way could be heard all over the ship. The sail had only a chance to slat two or three times, with force enough to almost carry the yard away and noise enough to drown the howling gale, before the men had hauled it up to the yard by its clewgarnets and buntlines. The Captain gave orders to put a reef in it before setting it again, which was quickly done.

That night was a stormy one, followed by a number of others, with the days thrown in; and it became necessary for the safety of the ship to heave her to. She was underwater most of the time, so deep loaded was she; and the strain on her was severe. When all was ready, a proper time was taken and the wheel put down.

The old girl came up with her head to the wind like a sea bird, and so nicely that she hardly took any water on board. When she had come to, all right, the main-topsail was clewed up and furled. On going below one could hardly believe such a gale was blowing, she rode the seas so quietly. She was the finest ship to lie in a gale of wind I ever was in. The harder it blew, the better she lay.

We lay to for three days and it was terrible to do so. The wind was fair for home; and instead of being some eight hundred miles nearer, we were hardly more than a hundred.

From then on we had favorable winds to abreast of Cape Horn, making the passage in thirty-five days. The weather was thick when we passed it, according to our reckoning, about twenty miles off. A person to fully appreciate the feeling that one has on rounding Cape Horn on the way home, when able to look at the compass and see that the ship is heading to the North, should be as we were, almost four years out. Not much was said by the crew about home (but one could see a quiet air of satisfaction amongst all hands) until we passed the Latitude of the Falkland Islands. Then every one of us youngsters broke loose. Jokes and fun were the order of the day.

On the fifteenth day after rounding the Cape, we were in the Latitude of 32° 50′ S., Long. 41° 20′; and so, as a little event had occurred the day after we left New Zealand that I think it well to mention, we will let the old ship go on her course with everything set on her from the royals to the lower studding-sails.

The morning after we left the Bay of Islands, two strange men were seen on deck amongst the crew. The Mate, seeing them, sung out, "Hollo! You strangers come aft here, let's have a look at you!" The two men came slowly walking aft to the Mate, who was standing at the break of the quarter-deck, gazing with astonishment at them as they approached; and well he might, for two more out-of-place-looking men would have been hard to find on any ship's deck.

One of the men was of large stature, a well-built man all round, with bright rosy cheeks, bright eyes and hair cut close to his scalp. On his head was an old straw hat with the rim half off and the crown partly loose. The upper half of his body was covered with an undershirt that had lost one sleeve and had numerous rents in it, here and there. No shoes were on his feet, but his pants had the stripe down each leg that is seen on such in the ranks of soldiers.

The other man was more slim built, quite tall, with bright sparkling black eyes and gentlemanly actions, and a very dark complexion (so much so that the boys nick-named him "Darkey"). On his short-cropped head was an old Scotch cap. The flannel shirt he had on was somewhat torn and his striped pants were not so much torn as his mate's, but his face, hands and clothing were covered with dirt and oil from crawling over the casks in the lower hold, where they had concealed themselves the night before the ship sailed.

On the Mate's asking who in the h—l they were and what they were doing on board this ship, the flannel-shirt fellow replied, "We are soldiers and want to go to the United States, so we ran away from our company."

The Mate told them to stand alongside of the mainmast until the Captain turned out for breakfast, when he would decide if they should be thrown overboard, put in double ropeyarns, keelhauled or kept on one cake of hard bread and all the small pieces they could eat a day, or not.

At six bells (7 A.M.) the Captain came on deck to take a

turn or two before sitting down to breakfast, and the Mate called his attention to the two stowaways standing by the mainmast, half-frightened out of their wits from the vague threats used by the Mate. As the Captain approached them, they came to as erect a position as the rolling of the ship would permit and gave him a military salute, which he answered by saying: "None of that lubberly work here! And may I ask you gentlemen, in the best and most kind, condescending, polite and humble manner, what in the name of all the little Devils in and out of hell do you mean by coming on board this ship in such a manner? And be damned to you for two lobster-backed, blue-and-white-spotted sons of sea cooks! I would be doing the proper thing by you to put you two sons of guns in the mizzin rigging and give you each four dozen." By this time the two half-famished, ragged and dirty beggars were almost frightened out of their wits.

The Captain took a turn back and forth on the quarter-deck. Then, stopping in front of them with a black and ugly look, he told them the ship was rendered liable to the English government in taking away soldiers from a garrison; and here we were some two hundred miles away from port. "To turn the ship back to land you might delay us a week, and be damned to you," said he.

The poor fellows, on hearing he might turn back with them, almost went down on their knees and begged him not to take them back, which none of us thought he meant to do. After a few more words of the like sort to them, he told them to go forward and get breakfast and to come aft when that was finished, which they did. He then gave them some clothes from the slop chest, likewise a pair of blankets, and they were set to work with the other men.

For green hands, they soon became quite useful. The elder of the two, by the name of Lowrie, had a narrow escape from death one night, though, when reefing the main-topsail. His escape was remarkable. We had been running with a heavy wind about four points on our starboard quarter under double-reefed fore- and single-reefed main-topsails, mizzin topsail furled, fore and main courses and jib, when it became necessary to put another reef in the main-topsail. The officer who had the deck should have hauled up the mainsail and rounded in on the weather-main and main-topsail braces, so as to spill the wind out of the sail in case he did not want to luff the ship off her course to do so. But no doubt, as he had so many men in his watch that he could depend on reefing the sail as the ship was running, he did not take the precaution he should have done.

The runaway soldier, Lowrie, was amongst the men who lay aloft on the yard to help reef the sail. Not understanding how to save himself, like the experienced sailors, when the sail, swelling up by the wind, presses upwards, he was caught and pressed backwards inch by inch, as the sail rose up forward of the yard like a bladder that is being blown up, until his fingers were forced from their hold on the jackstay and he was sent tumbling from the yard towards the raging sea some fifty or sixty feet below.

But, as he was in the starboard yard arm of the topsail about halfway out, directly over the weather main sheet that had been hauled taut from the tack that was boarded in the starboard cleat, when he fell, as good luck would have it, he struck the main sheet and his weight caused it to spring down; and when it rebounded, it threw him inside the bulwark on deck. There he was shortly after picked up insensible, but came to in a short time and was taken below. However, he had hardly recovered from the effects of the fall when we arrived home.

He could have been thrown outboard just as well as inboard, and if he had missed the sheet he would never more have been heard of. The ship was then going ten or twelve knots through the water and could not have been brought to quick enough to lower a boat and save him, even if the night was not so stormy.

The other soldier, by the time our ship arrived home, had become something of a sailor and could have shipped as an ordinary seaman with a show of some success. This man was something of a bully, and was inclined to be overbearing amongst the men, who had been taught, the whole voyage long, that fighting and quarreling was not healthy on board the ship. Of course, Darkey could not understand why the men did not resent his interferences; and no doubt he must have thought the "blasted Yankee sailors" had no fight in them. If so, he never was more mistaken in his life.

One of the best men and fighters amongst our men forward was a man who pulled the midship oar in our boat. His name was Matthews; and he had been on the Erie Canal from a boy old enough to ride a horse, until he got picked up by some runner for a shipping office in New York City and sent to New Bedford, with some others of the same stamp, where they shipped on board our ship.

One day, when we were North of the Line some degrees, Matthews was helping me clean the irons and lances that belonged to our boat. (They are always supposed to be scoured once a week.) While at work, he told me quite a number of things Darkey had said and done that were hard to put up with. From what he told me, one could plainly see that the fellow would be a better shipmate if someone would give him a good sound drubbing. Of course, the men could report him to the Captain, but none of them liked to, as that appeared beneath them. The fellow, he told me, had said he did not believe there was a man in the forecastle that had the pluck to fight. He believed Yankees as a rule were cowards, and if he could only have someone to knock off a little of the rust that was on him before he got to New

Bedford, it would do him good, as he then perhaps could find, after his arrival there, some man who might stand before him; and a lot of such vain talk.

It made me so wild that I told Matthews, the next chance he got, to give the beggar one or two licks where he thought they would do the most good. That night, in our watch on deck, I told the 2d Mate what Matthews had told me, and he said that if Matthews punched him he would not see anything of it unless he was obliged to; and that he would tell the 3d Mate, so he would look some other way if he happened to be around when it happened.

A day or so after my conversation with Matthews, the Captain and Mate were sitting away aft on the poop deck, taking the sun to determine the Latitude, and the 3d Mate and I were standing on the main hatch, when I heard a noise forward. Looking under the foot of the mainsail, I saw that Matthews and Darkey were making the rust fly. The 3d Mate, attracted by my movements, took a look with me just in time to see Darkey's heels disappear over the windlass, after his head that Matthews had knocked over it first. The 3d Mate quickly raised up, before any one should notice him, and started aft to the wheel, muttering, "I guess the rust is not so thick on the son of a gun as it was before."

As the foot of the mainsail and tryworks hid us pretty well from being seen from forward, no one could have an idea that the 3d Mate had omitted discipline in not interfering. Looking around, I could see that the Captain and Mate had observed nothing, so, making an excuse of looking at the jibs, I walked to the bows.

On the curve of the windlass bitts that hold the windlass from going forward sat poor Darkey, holding his eyes with both hands, with the blood running from his nose. One of the men brought a bucket of water to wash him as I turned to go aft. Matthews, with no sign of a scratch on him, was leaning against the fife rail to the foremast with his arms folded. Giving him a wink as I passed him, which he answered with a smile, I told the 3d Mate, when I met him, that I thought Darkey had rust enough off of him to last until the voyage ended. "How is this going to come out?" was the question I put to the 3d Mate.

"Well," he said, "when the 2d Mate comes down from the masthead, I will tell him the rust is pretty well off our soldier friend, and at dinner we will tell the Old Man all about what that fellow has been up to since rounding the Horn. And, as none of us aft has seen the fight, the Old Man, who enjoys a good thing when he can, no doubt will not take any notice of it." And he did not, even when Darkey was standing at the wheel, facing him, with the most lovely pair of black eyes you ever saw. A more peaceful shipmate than Darkey, for the rest of the voyage, could not be found.

THE LATITUDE we were in now was far enough from the Equator to soften the weather, and every pleasant day men were kept busy from morning until near sundown setting up riggin', putting on new rattlins, taking off the old ones and, when not too much worn, putting them on again, like the new, straight to a line with the upper set of deadeyes; the officer of the deck watching the men to see this done with exactness, and each hitch around the shrouds drawn properly up, the seizings neatly passed, and the ends properly fastened.

Tar buckets, slush shoes and slush buckets, serving mallets and serving boards, marlinspikes, fids, heavers, spunyarn, all kinds of seizing stuff from the size of twine to as large as a man's little finger, were in constant demand. By such time as we arrived in the Latitude of 25° North of the Line, the ship's riggin' was in perfect order, and, with the coat of paint on her yards and hull that we put on her, she had more the appearance of a ship just sailed from home than she did of one that was just off a four-years' voyage.

Gulf weed had been seen floating by the ship for some days, and it made me think of the encouragement it afforded to Old Christopher in his first efforts to discover new lands. I do not know as it affords others so much pleasure as it does me to see this stuff floating on the water when returning from a long voyage. It has such a charm for me that I have watched it for hours. At times, when my watch was below, I have got over the side into the main channels with my leg hooked around a chain plate, having a small net at the end of a long stick, and have dipped up bunches of this weed, in which would be found small crabs and fish of queer shapes and colors, some of which I have never seen anywhere else. Perhaps it is known where this weed grows, but I have failed in my efforts to find out. No one who has been shipmate with me could answer the question satisfactorily.

A day or so after painting ship, we saw a bark running down before the wind, and as she evidently wished to speak us, our mainsail was hauled up and the mainyard hove aback. As she approached us, everyone was commenting on her looks, with such exclamations as, "Look at her square yards," "Look at her sharp bows," look at this and look at that, until she swept by across our stern, when almost everyone broke out like one voice: "How beautiful!" Her sharp bows cut through the water, hardly turning a roll of foam the size of a necktie from them, and this almost disappeared as it slipped under her counters, that could carry no dead water there. Her sides were painted in fancy colors like a yacht.

When within hail, our Captain raised his speaking trumpet to his lips, and bellowed out, "Bark ahoy!"

The answer returned was "Hallo! What ship is that?"

"Ship *Charles W. Morgan* of New Bedford," says our Captain. "What bark is that?"

"Bark *Sea Fox* of Westport. How long are you out?"

"Forty-eight months. How long are you out, and what success?" (Our Captain.)

"Two months, two hundred and fifty barrels of sperm oil." (Stranger.) "What success with you?"

"Shamed to say, eleven hundred and fifty barrels." (Our Captain.)

"You have had hard luck." (Stranger.) "Won't you come on board?"

"No, the wind is fair and we want to get home as soon as possible," said our Captain. They waved their trumpets to each other as the last salute, we braced forward our main-yard and set the studding-sails, after boarding the main tack. The bark hauled to on the opposite tack and was soon lost to view in the distance.

The appearance of this vessel was the subject of talk for hours afterwards. It was to all of us like a vision of beauty, as this was the first clipper ship any of us had seen. The building of them had commenced in its fullest extent while we were away, and their model, so different from the old style of square boxlike ships, made the contrast more striking. It is said that Donald McKay, who built ships of that description, used the lines of the canoes that the Indians on the Northwest coast of America have; and on seeing them, one can believe the report correct.

Sails were seen more frequently as we made North. A Spanish brig from Havana bound to Malaga came very near a collision with us one night. Our lookout on the bows reported a sail in sight heading on the wind to the E. N.E., we steering N.N.W., wind S.E. She was close on our port bow when first seen, and as we put our wheel up to pass astern of her, she also put her wheel to starboard and we had hard work to clear the lubberly, garlic-eating, vile-smelling stupid jackasses aboard of her, who must have all been asleep and knew not what to do, on being roused up on our hail of "Brig ahoy!"

About two hours after we had cleared the brig, the wind suddenly hauled out N.E. and blew great guns, with heavy thunder and sharp lightning, and a heavy downfall of rain that gave the watch a complete soaking while reducing sail to double-reefed topsails, jib and mainsail furled. This trouble lasted about twelve hours and then the wind changed to S.E. again. All sail was then set and away we went booming on our course again, with the current in our favor of at least four knots, as we were now in the Gulf Stream.

On the 25th day of May, 1853, we hove the ship to and got soundings in sixty-five fathoms of water. We supposed ourselves about fifty miles from Block Island, but as we had had no observations for some days, we could only judge our position by the soundings and dead reckoning. The wind was still from the South and East, the weather thick and rainy, so we kept a bright lookout and steered our course N.N.E. All hands were set at work tearing down the tryworks and throwing overboard the bricks and useless lumber from it. The trypots were turned bottomside up and lashed between the heavy iron knees, bolted to the deck, that hold the tryworks in position. The tryworks are always torn down at the end of each voyage and new ones built for the next.

We ran on our course some thirty or forty miles, the weather still remaining rainy and foggy. The ship was hove to under low sail, and at daylight the wind was very light, but the fog was so thick that some of the boys said, "The only way the flying jibboom can penetrate it will be for a man to open a hole in it with a marlinspike."

At six bells we went to breakfast, but not much was eaten by me. I was nervous, and soon went on deck again. Soundings were taken, on and off, during the night, and they placed the land some ten or fifteen miles away, so, if the weather would only clear a bit, the land would plainly be seen. But nothing could be seen a ship's length in any direction, and the dirty drip, drip, drip of the fog from the riggin' made it very uncomfortable for those on deck.

About four bells a light place in the clouds showed the sun was trying to break through. By eleven A.M. patches of blue sky could be seen here and there overhead. The Mate sung out for me to go aloft and see if anything could be seen over the fog as it still hung thick around the ship. Jumping into the main riggin', I went aloft to the main-royal yard. Taking a look around me, I could see two or three masts of vessels sticking out above the fog. They looked as though the hull and lower masts were submerged in a sea of smoke. Everything around was silent; but the slat of our topsails and the patter of the reef points, as the old ship rose and fell in a sluggish manner on a lazy ground swell, gave one a queer, weirdlike feeling, and thoughts of phantom ships and the Flying Dutchman. Some fifteen or twenty minutes elapsed, and then the fog came sweeping in around our mastheads, shutting out of sight everything beyond the ends of the royal yard.

The rest of the day was passed in laying aback with our mainyard to the mast, on different tacks, and all hands cussing at the fog that kept us from getting to our long-absent homes. By night it would have been hard to find a crew of men so totally cross in the whole Atlantic Ocean. The night watches were most uncomfortably passed, and, as soundings had become monotonous, none were taken. Daylight broke with the same tired wet look on all hands who had had the watches during the night.

When the Mate came on deck at four bells (six A.M.) he had a look on his face that would turn milk sour, and could hardly speak civil to the officer of the deck. The Captain came on deck just before the bell struck to call the watch for breakfast, and he was not a bit more pleasant. When breakfast was called, I went with the Captain and officers to it, as it was my turn (each boat-steerer takes his regular turn) at the first table.

After breakfast, where but little had been said by anyone, we went on deck. The 3d Mate went to the galley, took a coal of fire from the stove to light his pipe, and when he had it well lighted, went to the side to throw the coal overboard. In the act he suddenly removed his pipe from his mouth and stood for a minute with his head over the rail in the position of one listening. Happening to have my eyes on him at the time, I could not tell what it meant until he suddenly stepped back from the rail and sung out to the Captain: "We have a barnyard on the lee bow! I've heard a cock crow and a cow bellow."

"Hard up the wheel! Wear ship," the Captain sung out.

The wheel was put hard up, the yards squared in, none too quick either, for, as the ship fell off from the wind, a dark mass with a nasty line of breakers in front of it came crawling out to windward almost under our flying-jibboom end. If the officer had not heard the sounds when he did, we would have ended our voyage, and our ship too, on the east end of Long Island. As it was, we stirred up the sand with our keel but did no damage to the ship. There was considerable excitement around the decks for a short time.

By four bells (ten A.M.) the fog had got so thin that objects as large as a ship could be seen a mile away. All sail was set, as the wind now hauled out N.W., and by eight bells (twelve M.) the weather cleared off and the sun shone as bright as our faces now were at the sight of Block Island on our port bow, about five miles off, and the shores of Buzzards Bay plainly in view right ahead, which we were nearing fast. With a strong N.W. wind four points free, the old ship was tearing up a sheet of foam with her bows that made a roar that was delightful to hear, as she logged at least eleven or twelve knots. Lying well over, her braces taut, yards swaying and masts creaking, she seemed only too glad to make up for the many weary hours she had been keeping us away from homes and loved ones.

Early in the afternoon a pilot boat ran close to us, dropping his small boat. We luffed into the wind and soon had an old Sound pilot on board. Keeping off on our course again, we ran up to the land while he told us of many things new which had taken place since our leaving home. None were so startling as the fortunes that had been made in California; and on his naming some whom we had been to school with, as returning with wealth, the thought went through my head that they had not been through half what we had; yet how different the results. Most of the crew and officers of the ship were in debt, after four-years' service.

Leaving on our right some rocks called the Hen and Chickens, with Clark's Point Light open on our port bow, we steered into the harbor of New Bedford, passing Palmer's Island. The wind continued fair and we ran the ship into the mud, hard and fast, about twice her length from Robinson's Wharf.

I was sent with the boat to run a line to the wharf, that the ship might be hauled alongside of it when the tide rose. The rest of the men clewed up and furled the sails. The boat's crew, with me, pulled in to the wharf. When I threw up the end of the line I had brought with me, it was caught by a cousin of mine, who, with three or four other cousins and numbers of my acquaintances, had come down on the wharf to meet me.

Making the line fast, my cousin sung out for me to get out of the boat and come with him. I hardly liked to leave the ship that way, I told him. "Well," he said, "there is nothing else to do but furl the sails, until the tide rises, and the shipkeepers that the owners will send on board will haul the ship in to the wharf."

Before I got out of the boat I found out that Mother and

Sister were well; but, not knowing when the ship would arrive, they had not come to meet me, and were still at home in Maine. I told the boat's crew to take back the boat and tell the Mate that, meeting some of my friends, I had gone with them. When I got on the wharf, the boys surrounded me three deep, and we marched up the wharf talking and laughing like a lot of schoolboys.

The feeling of pleasure to once more be walking through the streets so familiar, cannot be expressed by me in words now. Many times I had dared not let my thoughts carry me away, as it seemed, when thinking over in the lone watches on deck at night, or swaying back and forth at the topgallant head when on the lookout for whales, that my chances of being killed by a whale or falling from aloft or tumbling overboard were as much for it as against it; and it often had appeared I never would have such pleasure again.

The first landfall the boys took me to was a barber shop. My hair was cut, whiskers trimmed, and a hot bath taken, and then to an infitters, or clothing store, where the boys saw me fitted in a suit of stylish clothes that did not require reefing in back, arms or legs. Letting the cutter of this establishment take my measure for another suit of clothes to be ready for me at the end of two days, as I wished to take the train in that time for home, we started for supper at the house of one of my aunts.

From the owner of the store I had drawn some money. This and the clothes had to be paid for when I settled my voyage, after the oil had been hoisted out on the wharf and tested, and the amount determined, some two or three weeks later. After stopping in New Bedford for two or three days, having a good time with my old schoolmates and cousins, who made quite a fuss over me, I then took the train for Boston on my way to Maine. The cars in which I rode made good time, no doubt about that, but, impatient as I felt, they could not go fast enough for me.

Arriving home the next day, and having Mother's arms around me, all sense of fear left that had hovered over me, that something might occur to prevent our ever meeting again. (Why for months I had such a misgiving, is hard to say.) When Mother and Sister had got through with me, the old Colonel, my stepfather, took me in hand. Colonel Dunn seemed pleased to see me and thought I had improved much in appearance; and on my informing him of the numbers of offers that had been made me to go as officer of different ships, "I feel proud of your standing," he said.

After a week or so at home, I went down to New Bedford and settled up my voyage, which amounted to about four hundred dollars. It had cost me, for my outfit and what I had drawn on the voyage, about $200.00; so I had about $200.00 to show for my four-years' work. Such is Fortune. "The sweet little Cherub who sits up aloft, to keep watch of the life of poor Jack," does not attend to the duty over and above well. Settling up my bills in New Bedford, I returned home.

During the time I was settling my voyage, quite a number of the shipowners offered me good chances to ship as officer in their ships; but I refused to bind myself to any, as I had promised Mother, before leaving to settle my voyage, that I would not engage myself for sea again, right away.

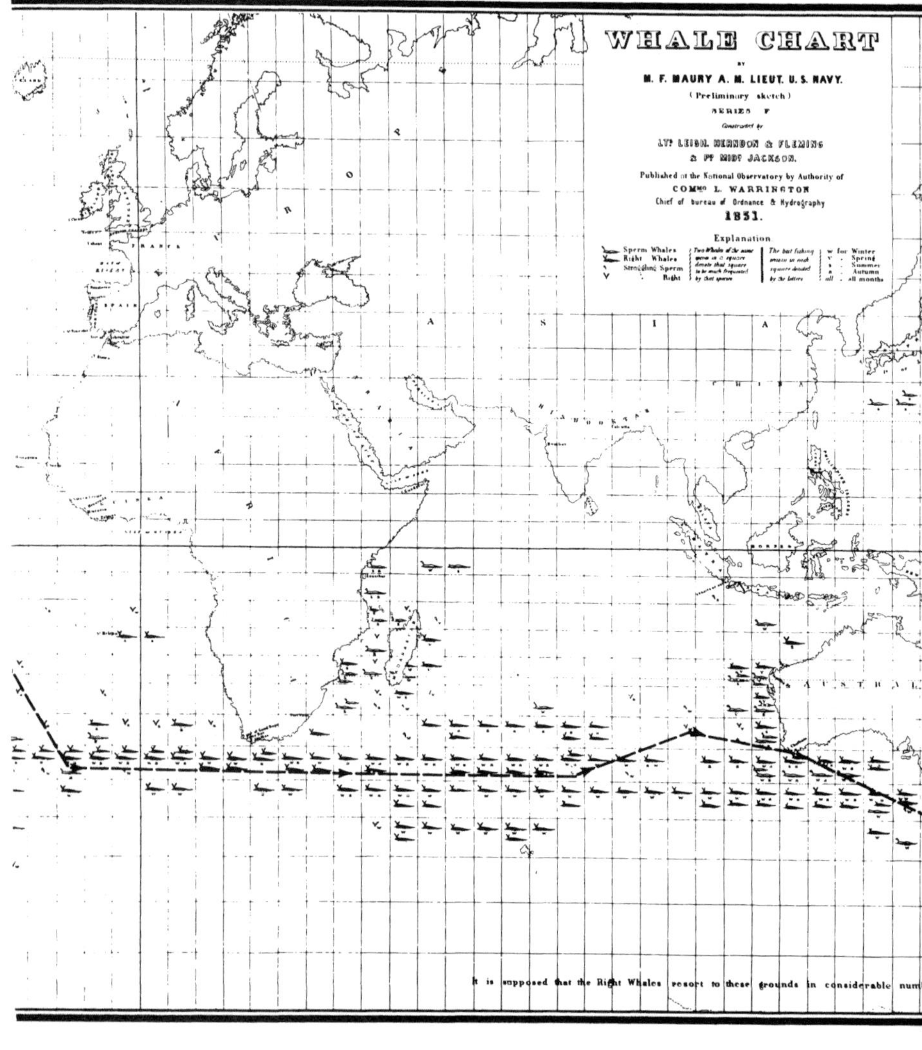
WHALE CHART
BY
M. F. MAURY A. M. LIEUT. U. S. NAVY.
(Preliminary sketch)
SERIES F
Constructed by
LT? LEIGH. HERNDON & FLEMING
& P? MID? JACKSON.
Published at the National Observatory by Authority of
COMMO L. WARRINGTON
Chief of bureau of Ordnance & Hydrography
1851.
Explanation
Sperm Whales
Right Whales
Straggling Sperm
" Right
Two Whales of the same genus in a square denote that square to be much frequented by that species
The best fishing season in each square denoted by the letters
w for Winter
v - Spring
s - Summer
a - Autumn
all - all months
A S I A
It is supposed that the Right Whales resort to these grounds in considerable num

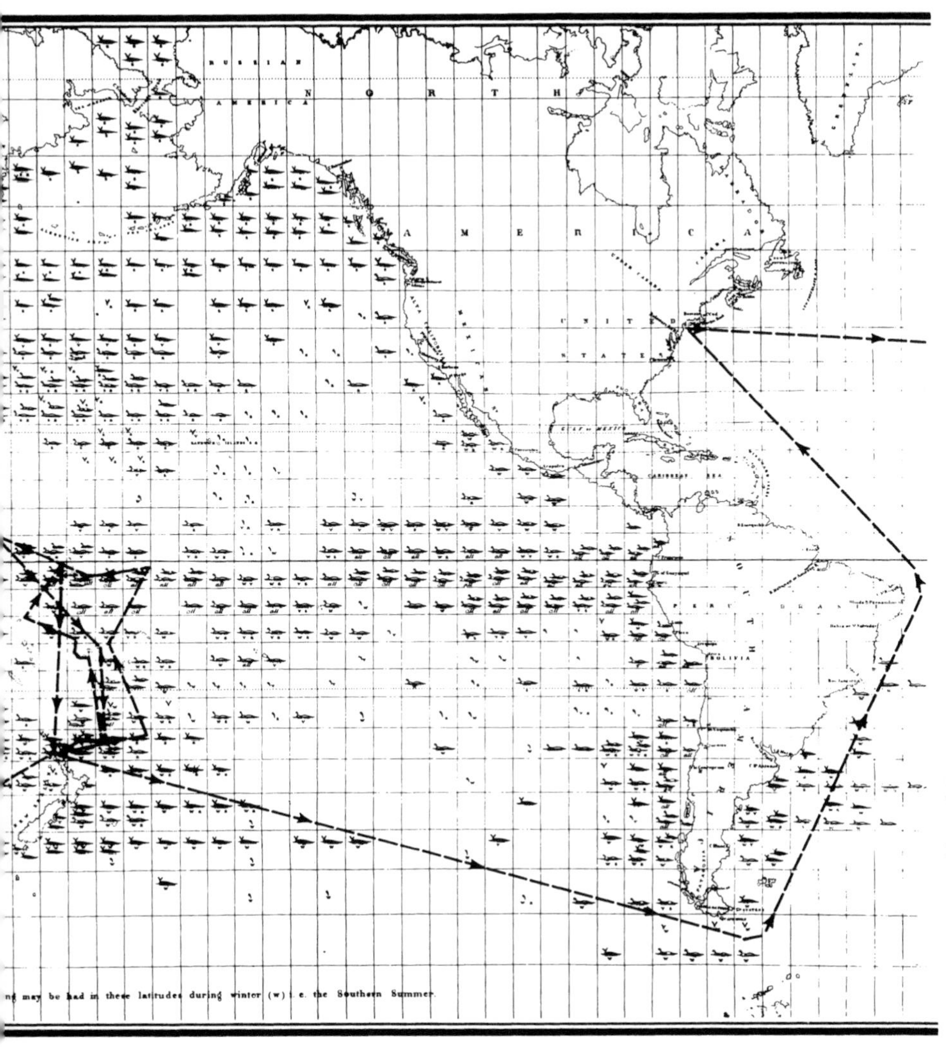

The approximate track of the 1849–53 voyage of the *Charles W. Morgan* is here superimposed on Lieutenant Matthew Fontaine Maury's whale chart, published by the U.S. Naval Observatory and Hydrographical Office in 1851.

Also published by Mystic Seaport

"She Was a Sister Sailor" *The Whaling Journals of Mary Brewster, 1845-1851*
Joan Druett, editor

Defying convention, Mary Brewster of Stonington, Connecticut, accompanied her whaling captain husband on two voyages to the Pacific Ocean, becoming the first American woman to sail into the Western Arctic. Joan Druett, a well-known New Zealand novelist, has fully annotated this fascinating daily record of a woman's life in a man's world at sea, adding numerous quotes from other whaling women, lively introductory essays to give context to the journals, and a list of all the known women who went whaling. Now in Mystic Seaport's G.W. Blunt White Library, Mary Brewster's journals are here published for the first time. As the most complete account of the female experience at sea, this volume will be of great interest to both scholars and enthusiasts of whaling and maritime history, Pacific history, and women's history. *"She Was a Sister Sailor"* was recognized by the North American Society for Oceanic History as the best non-naval book of nautical history published in 1992.

(1992) 7-3/8"x 10-7/8", 480 pages, 63 illustrations, appendices, annotated bibliography, index.

ISBN 0-913372-60-9 (cloth) $39.95

The *Charles W. Morgan*
John F. Leavitt

Out of print for several years, Mystic Seaport's illustrated history of its most significant vessel has been revised and redesigned. This new edition augments John F. Leavitt's text and illustrations with more than 80 photos, most of which document the *Morgan*'s crew at work on the whaling grounds, along with 20 drawings, 10 of them studies of the vessel's structural details by Kathy Bray. There are also five 8" x 16" pullout plans bound into the book for reference by modelmakers – sail and rigging plan, deck and bulwark plan with details of gear, construction plan, plan of square sails and their rigging, and a cutaway view by Roger Hambidge, museum shipwright, who also contributed an Afterword that discusses the *Morgan*'s refloating and restoration between 1973 and 1990.

(1998) 8" x 9", 124 pages, 108 illustrations, appendices, glossary, reading list, index.

ISBN 0-913372-10-2 (paper) $24.95

America and the Sea *A Maritime History*

William M. Fowler, Jr.
Andrew W. German
John B. Hattendorf
Benjamin W. Labaree
Jeffrey J. Safford
Edward W. Sloan

America and the Sea: A Maritime History is the most comprehensive maritime history of the United States available today. Spanning the centuries from Native American and Viking maritime activities before Columbus through today's maritime enterprise, the text provides a new history of the U.S. from the fundamental perspective of the sea that surrounds it and the rivers and lakes that link its vast interior to the seacoast. *America and the Sea* has been gracefully written by six prominent maritime-history scholars whose individual areas of research and teaching range from labor to technology, from the fisheries to the U.S. Navy. Their text incorporates considerations of art, literature and poetry along with discussions of the economic, political, diplomatic and technological foundations of American maritime history. Their narrative is punctuated and augmented with quotations from period documents, and with brief essays by younger scholars that add insight and expand on the human dimensions of America's relationship with the sea. This rich, complex story is told not only in words but with 71 color images, many of them paintings from museums around the world, with 289 archival photographs and drawings in black and white, and with 10 full color maps.

(1998) 9" x 12", 694 pages, 360 photographs
and drawings, maps, appendix, bibliography, index.

ISBN 0-913372-81-1 (cloth) $65.00

"A model of popular scholarship, and clearly the definitive work on the subject."
– Kirkus Reviews

"Abundantly illustrated with photos, graphs, and full-color maps, this is essential for all libraries."
– Library Journal